Engineering Materials

This series provides topical information on innovative, structural and functional materials and composites with applications in optical, electrical, mechanical, civil, aeronautical, medical, bio- and nano-engineering. The individual volumes are complete, comprehensive monographs covering the structure, properties, manufacturing process and applications of these materials. This multidisciplinary series is devoted to professionals, students and all those interested in the latest developments in the Materials Science field, that look for a carefully selected collection of high quality review articles on their respective field of expertise.

Indexed at Compendex (2021) and Scopus (2022)

Wendy Triadji Nugroho · Yu Dong ·
Alokesh Pramanik

Thermoplastic Polyurethane (TPU)/Halloysite Nanotube (HNT) Nanocomposites

Additively Manufactured Using Fused Deposition Modelling (FDM)

Springer

Wendy Triadji Nugroho
Engineering Department
Politeknik Negeri Jember
Jember, East Java, Indonesia

Yu Dong
School of Civil and Mechanical
Engineering
Curtin University
Perth, WA, Australia

Alokesh Pramanik
School of Civil and Mechanical
Engineering
Curtin University
Perth, WA, Australia

ISSN 1612-1317 ISSN 1868-1212 (electronic)
Engineering Materials
ISBN 978-981-95-0540-1 ISBN 978-981-95-0541-8 (eBook)
https://doi.org/10.1007/978-981-95-0541-8

This Springer imprint is published by the registered company Springer Nature Singapore Pte Ltd.
The registered company address is: 152 Beach Road, #21-01/04 Gateway East, Singapore 189721, Singapore

Preface

Additive manufacturing (AM), known as 3D printing, has drawn great attention due to its ability to fabricate structures with complex geometry, design freedom and reduction of material waste from manufacturing processes. One of the most popular techniques in AM is fused deposition modelling (FDM), which uses filaments as the feedstock material and builds the parts based on melt extrusion method. Besides, FDM is capable of processing various materials including polymers and metals. This technique is user-friendly, and thus can be classified as a typical low-cost manufacturing technique.

Due to good biocompatibility and flexibility, polyurethane (PU) has been widely used in focused areas such as biomedical and electronic fields. PU is created via a chemical reaction between isocyanate functional groups and hydroxyl groups of polyols, resulting in both soft and hard segments. The soft segments form continuous matrices for low-temperature flexibility, while hard segments tend to be self-assembled into the domains through crosslinks. The physical crosslinks of PU are generally reversible in nature, which means that above its glass transition temperature T_g, both soft and hard segments become soft in the formation of a homogeneous mixture. It is generally ascribed to PU morphological structures achieved by corresponding chemical structures and processing conditions. Among PU family, some PUs with excellent shape memory effect (SME) are known as shape memory PU (SMPU). This type of materials can memorise and regain its initial shape after being physically deformed by imposing external stimuli such as heat, light, electrical and magnetic fields. They can meet special requirements of tissue engineering as biodegradable scaffolds, wearable devices, sensors and flexible circuits.

However, PU has low mechanical properties in terms of mechanical strength, stiffness and impact resistance so that reinforcement materials such as halloysite nanotubes (HNTs) are utilised to enhance resulting material properties. In contrast with other nanofillers such as carbon nanotubes (CNTs) that have been widely used as reinforcements, HNTs in kaolin family are naturally available with high cost-effectiveness. Meanwhile, Both HNTs and CNTs have very similar hollow tubular structures, which make HNTs a potential material candidate as relatively cheap nanofillers.

This book consists of three major research components, namely preliminary, main and application-oriented studies. The preliminary study can be divided into two sections. The first section investigates the effects of 3D printing parameters in terms of nozzle temperature, print speed, infill density and layer height on dimensional accuracy and surface quality of 3D printed samples made from neat PU by utilising L_9 orthogonal array (OA) and Pareto analysis of variance (ANOVA). Taguchi design of experiments (DoEs) and Pareto ANOVA were employed to identify significant processing parameters and optimum factor-level combinations in response to dimensional accuracy and surface quality of PU. Meanwhile, the second section aims to evaluate the influence of HNT content and 3D printing parameters (i.e. nozzle temperature, print speed, infill density and layer height) on dimensional accuracy and surface roughness of 3D printed PU/HNT composites using L_{18} OA and Pareto ANOVA. Preliminary study was conducted to gain adequate knowledge in relation to AM technologies and the particular feasibility of FDM as the selected 3D printing process. On the other hand, main study implemented L_{18} OA and Pareto ANOVA to assess the impacts of HNT content and 3D printing parameters on tensile properties (i.e. tensile strength, tensile modulus and elongation at break), hardness and abrasion resistance of PU/HNT nanocomposites manufactured by FDM technique. Thermal properties, including thermal stability, glass transition temperature T_g, crystallisation temperature T_c, melting temperature T_m and degree of crystallinity χ_c, were also discussed. Finally, application-oriented study relying on the optimum factor-level combination focused on the effect of HNT content on compressive strength, compressive modulus, specific energy absorption (SEA), Poisson's ratio (PR), shape fixity ratio, shape recovery ratio and recovery time of PU/HNT nanocomposite samples. Besides, this study also revealed the influence of HNT content on the crack propagation of unit cells in RE samples induced by compressive tests.

In this book, Chap. 1 focuses on comprehensive review of popular additive manufacturing used for polymers and composites including binder jetting (BJ), directed energy deposition (DED), material jetting (MJ), powder bed fusion (PBF), sheet lamination (SL), vat photopolymerisation (VP) and FDM along with individually used 3D printers and their specifications. Chapter 2 covers five fundamental steps in additive manufacturing in range of creation of 3D computer-aided design (CAD) model, conversion into standard tessellation language (STL) format, setup of printing parameters, additive manufacturing processes and removal and clean-up processes. Chapter 3 highlights the fabrication of PU composite structures using 3D printing that involves purification, dehumidification and mixing processes in the preparation step. Chapter 4 reviews the manufacture of re-entrant (RE) structures made from PU composites using 3D printing based on four different types of structural models, namely conventional honeycomb, RE honeycomb, tetrachiral and arrowhead structures, along with different manufacturing techniques, modelling approaches, as well as properties of 3D printed RE structures. Chapter 5 introduces 4D printing technologies specially used for PU composites, resulting properties of 4D printed structures, as well as associated SME, external stimuli and mechanisms. Chapter 6 discusses about detailed research methodologies and analytical techniques, which are not limited to preparation of material samples, experimental characterisation and parametric optimisation using DoEs and Pareto ANOVA. Chapter 7 covers dimensional accuracy in terms of full length, grip section width, thickness, circularity and cylindricity, as well as surface quality with respect to surface roughness and surface waviness for PolyFlex^{TM} PU95 dog-bone samples according to ASTM D638 (Type V) standard. Chapter 8 reports the manufacture of polyurethane (PU)/HNT nanocomposites with optimum factor-level combinations via DoEs according to L_{18} orthogonal array with the neglect of factorial interaction. Chapter 9 concentrates on the influence of both nanofillers and 3D printing parameters on mechanical and thermal properties of 3D printed PU/HNT nanocomposites according to pre-determined optimum factor-level combinations. Chapter 10 evaluates the effects of different HNT content on shape memory behaviour and compressive properties of PU/HNT nanocomposites manufactured via FDM. Strip-like and RE structures were manufactured based on PU (MM4520 grade) and HNTs with the combination of solution casting, melt extrusion and 3D printing. Finally, Chap. 11 elaborates potential applications of PU/HNT nanocomposites via 3D printing in range of six different areas including biomedical engineering, electronic and antistatic devices, thermal insulation, sports equipment and garments for future product development.

The first author Wendy Triadji Nugroho acknowledged continued financial support from both the Indonesia Endowment Fund for Education, known as Lembaga Pengelola Dana Pendidikan (LPDP), under the Indonesian Ministry of Finance, and School of Civil and Mechanical Engineering at Curtin University. We are indebted to laboratory technical staff Rob J. Walker, Vince Skyrme, Melina Miralles, Matt Aylmore, Leon Forgus, Darren Isaac, Ming Lim, Luke Mickan, Elaine Miller, Aaron Dodd, Yu Long, Marisa Yookhong and Peter Chapman at Curtin University. In addition, we would also like to give a special mentioning to editorial team at Springer Nature including Loyola D'Silva, Yuva Shree Sankaran, Praveenkumar Boominathan and Diya Ma to make this published book a great success.

We sincerely welcome valuable comments and feedback from eminent researchers, scientists and materials engineers to improve the overall quality of this book. We anticipate that this book can become a good technical resource in the cutting-edge field of additively manufactured polymer nanocomposites.

Jember, Indonesia — Wendy Triadji Nugroho
Perth, Australia — Yu Dong
Perth, Australia — Alokesh Pramanik

Competing Interests The authors have no competing interests to declare that are relevant to the content of this manuscript.

Contents

Chapter 1
Popular Additive Manufacturing for Polymers and Composites

Abstract This chapter focuses on a fundamental knowledge of additive manufacturing, also known as 3D printing, which is implemented for manufacturing 3D printed parts based on polymers and composites. Seven 3D printing methods are discussed, which consist of binder jetting (BJ), directed energy deposition (DED), material jetting (MJ), powder bed fusion (PBF), sheet lamination (SL), vat photopolymerisation (VP) and fused deposition modelling (FDM). Individual 3D printers for different methods along with their specifications are also mentioned in this chapter to provide comprehensive equipment-based information.

Keywords Additive manufacturing · Binder jetting · Directed energy deposition · Material jetting · Powder bed fusion · Sheet lamination · Vat photopolymerisation · Fused deposition modelling

Additive manufacturing (AM), also known as 3D printing, refers to a cutting-edge manufacturing technology to combine various materials for fabricating physical objects according to 3D computed aided design (CAD) model data [1]. The ability of 3D printing to build an object with complex geometry with open cavities like lattice structures makes it widely used for a range of applications, not only in engineering industries but also in other fields like education, architecture, toys, entertainment, cartography and medicine. 3D printing has the capability to implement on-demand manufacturing of replacement objects. Besides, those objects also can be printed by local distributors in remote regions without delivery constraints [2].

The primary difference between additive manufacturing and subtractive manufacturing is that AM creates an object by depositing material layer by layer so that it can minimise material waste. On the other hand, subtractive manufacturing involves material removal such as turning, milling, drilling, grinding, cutting and boring. However, AM will not completely replace other manufacturing methods (e.g. subtractive manufacturing, injection, forging, etc.) due to typical issues encountered with surface finish, fabrication speed and mechanical strength of 3D printed objects [2]. According to the International Organisation for Standardisation (ISO) and the American Society for Testing and Materials (ASTM), 3D printing techniques are divided

W. T. Nugroho et al., *Thermoplastic Polyurethane (TPU)/Halloysite Nanotube (HNT) Nanocomposites*, Engineering Materials, https://doi.org/10.1007/978-981-95-0541-8_1

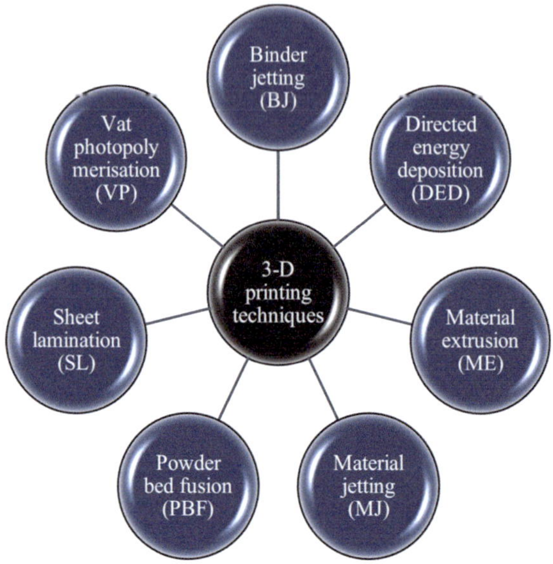

Fig. 1.1 Seven classifications of 3D printing techniques [4]

into seven classifications [3], as illustrated in Fig. 1.1. Detail information regarding such techniques is available in Sects. 1.1–1.7.

1.1 Binder Jetting

Binder jetting (BJ) utilises powder particles, which are bonded with liquid binder agents as feedstock materials. Mostly, powder particles are placed in a powder bed, and only a small portion is located in the print head. Afterwards, liquid binder agents are ejected from nozzle tips as the droplets (typical diameter: 80 μm) [5]. When touching powder particles on the powder bed, such droplets then turn into spherical agglomerates and bond previously printed layers (see Fig. 1.2).

In general, BJ print heads possess several nozzles, thus enabling a few 1D parallel ways for patterning. This 3D printing process can be scaled economically and categorised as scalable and line-wise patterning by only appending the number of nozzles in print heads [4]. This unique characteristic allows BJ to build a part with a high deposition rate and minimum cost.

Powder particles used by BJ comprise metals (e.g. titanium, copper and steel) ceramics (e.g. carbon, carbide, alumina, silicon and sands), metal composites (e.g. tungsten/bronze composites) and ceramic/metal composites (e.g. boron carbide/aluminium and silicon carbide/silicon composites). Feedstock materials in powder form are filled into a hopper in BJ process, and a wiper then spreads such powders across a build platform. A head of inkjet printer selectively spouts some fine droplets of the liquid binder onto a bed platform in order to glue powder particles very well.

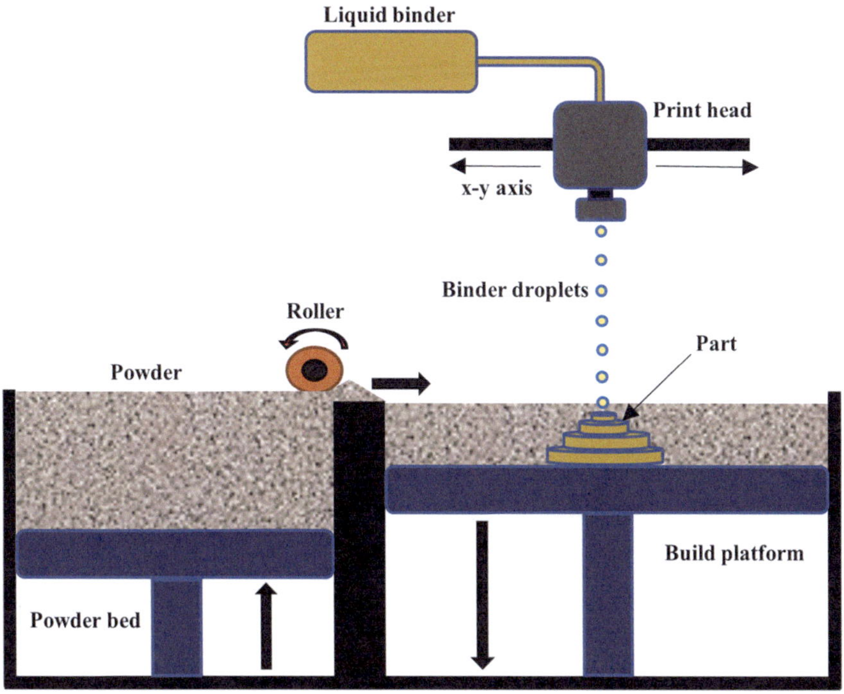

Fig. 1.2 Schematic diagram of binder jetting [4]

This process is repeated several times to produce an entire part. The resulting part is called a "green part", which has not been a dense structure yet since individual particles have not physically bonded with each other [6]. Accordingly, a two-step post-processing is required to improve both density and strength of 3D printed parts. The first step involves a curing process, which moves the build volume and places it in an oven. The green part is then cured to enhance its mechanical strength. Once this step is finished, the part is removed from a powder bed and sprayed with compressed air in order to clean up any loose powders [7]. The second step implements a sintering process, in which powder particles are heated at high temperatures in an inert environment, which is followed by a released binder to increase part density. A hot isostatic pressing (HIP) can be added to produce a highly dense part in order to meet the end-users' specific requirements [7].

Table 1.1 presents five BJ machines, consisting of VX200 [8], Innovent$^+$ [9], DM P2500 [10], M-Flex [11] and S-Max [11] along with their manufacturing specifications. Among them, Innovent$^+$, M-Flex and S-Max are supplied by ExOne Company, while DM P2500 and VX200 are manufactured by Digital Metal and Voxeljet respectively. S-Max has the largest build volume, as opposed to other 3D printers. Besides, it also can manufacture a part much faster than others with 100,000 cc/h. Nonetheless, the print resolution of S-Max is very low at 100 μm. On the other hand, Innovent$^+$

Table 1.1 BJ machines and their manufacturing specifications

Variable	VX200	Innovent+	DM P2500	M-Flex	S-Max
Build volume (mm)	300 × 200 × 150	160 × 65 × 65	203 × 180 × 69	400 × 250 × 250	1800 × 1000 × 700
Maximum build rate	14 mm/h	166 cc/h	100 cc/h	1600 cc/h	100,000 cc/h
Layer height (μm)	300	30–200	> 150	50–200	260–380
Print resolution (μm)	300 (300 dpi)	30	35	50	100
Printable material	Sand/poly (methyl methacrylate) (PMMA)	Metals, ceramics, composites	Metals (e.g. stainless steel, Ti6Al4V)	Metal, ceramics and sand	Sand
Machine dimension (mm)	1700 × 90 × 1500	1203 × 1016 × 1434	2900 × 1000 × 1700	1675 × 1400 × 1855	10,400 × 3520 × 2860

with the lowest layer height can process a variety of materials such as ceramics, metals and composites despite the smallest build volume.

1.2 Directed Energy Deposition

Directed energy deposition (DED) typically uses an electron or laser beam as a heat source and directs it into focused areas in order to heat and simultaneously melt both feedstock materials and the substrate, as illustrated in Fig. 1.3 [3]. A track resulting from solidified materials is created by the route of DED head, while the layers are created by adjacent material lines. Such a 3D printing method involves an inert gas to carry the powders to the substrate in order to prevent the deposit from oxidation.

The feedstock materials in DED process usually come from metals such as titanium, nickel and steel. Nevertheless, with current technological advances, DED also enables to process composites such as tungsten carbide (WC), titanium carbide (TiC) and continuous carbon fibre-reinforced thermoplastics (CFRTPs). The two-axis or three-axis system in DED comprises a static base plate, in which the nozzle moves up to form a layer. In a four-axis or five-axis system, both the nozzle and the base run at the same time, which are independent of each other [6]. A flat starting surface is not obligatory for DED process. As such, it can add new materials to part surfaces or deposit them on damaged parts for repair.

DED is capable of processing feedstock materials in powder and wire forms with both their advantages and disadvantages. The powder form can easily change the orientation between layers since the presence of excessive powder flow induces

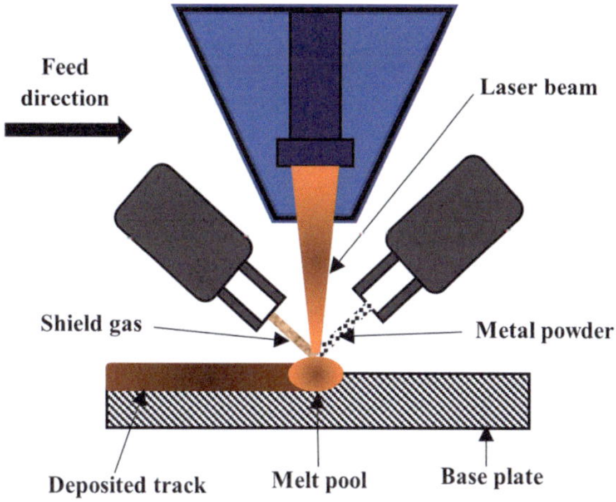

Fig. 1.3 Schematic diagram of directed energy deposition [4]

dynamic levelling on the melt pool and deposit the thickness at each deposited layer area. On the other hand, not all powders can be captured in the melt pool. Consequently, the quantity of such powders decreases during the manufacturing process. As such, excessive powders are required to maintain the availability of feedstock materials. In case that powder recycling is in demand, an excess of powders need to be recaptured in a clean condition, which is not always easy to perform.

The advantage of wire form is that the efficiency of feedstock capture is close to 100% since the amount of feeder wire is almost the same as the deposit volume [12]. Furthermore, the wire can be employed to create parts with simple or block geometries without many transitions and also to coat component surfaces. Meanwhile, for complex, dense and large models, processing parameters like wire diameter, layer thickness and wire feed rate must be carefully controlled in order to produce desired deposit size and shape.

Table 1.2 presents four different 3D printers based on DED technique, namely μPrinter [13], FormAlloy L-Series [14], LENS CS 150 [15] and LENS MTS 500 [15], which are manufactured by Additec, FormAlloy and Optomec respectively. μPrinter can deposit both wires and powders into complex geometries with the power consumption of only 600 W. However, this type of 3D printer possesses small build volume with the lowest accuracy. FormAlloy L-Series is capable of performing an auto-set layer thickness, as well as auto-correct errors to produce a part without dimensional defects. This printer has better accuracy than μPrinter though worse than Optomec machines. LENS CS 150 and LENS MTS 500 produced by Optomec can build a part with high accuracy up to 0.014 and 0.005 mm respectively. Furthermore, LENS MTS 500 possesses the largest build volume and utilises hybrid additive and subtractive metal processing that combines laser engineered net shaping (LENS) and computer numerical control (CNC). Accordingly, this combination may result in desired parts with high accuracy and excellent surface quality.

Even though DED machines are classified as high-cost machines due to the application of laser technology and powdered metals to manufacture the parts, such

Table 1.2 DED machines and their specifications

Variable	μPrinter	FormAlloy L-Series	LENS CS 150	LENS MTS 500
Build volume (mm)	160 × 120 × 450	200 × 200 × 200	150 × 150 × 150	350 × 325 × 500
Laser power (W)	600	Up to 8000	400	500–2000
Accuracy (mm)	9	0.5–8	0.014	0.005
Printable material	Metals (any weldable materials)	Metals (e.g. nickel, iron, titanium, cobalt, copper)	Metal (e.g. stainless steel, Inconel)	Metal (e.g. stainless steel, Inconel, tungsten carbide, Stellite, Hastelloy, tool steel)

machines can perform rapid prototyping, repairing and remanufacturing of legacy parts from metal materials.

1.3 Material Jetting

Material jetting (MJ) builds a part using a common method similar to a typical 2D inkjet printer. This kind of technique utilises ink as its feedstock material. The ink is commonly a photopolymer resin, which is sensitive to ultraviolet (UV) light. MJ allows to process many kinds of feedstock materials since this technique possesses some print heads that eject such materials out from nozzle tips in the form of picoliter droplets to targeted areas (see Fig. 1.4). Those deposited droplets are then solidified by UV curing lamp to manufacture a part. MJ is capable of creating a part in a high *z*-direction resolution with the layer thickness up to 16 μm [16].

The ink should be heated up to 30–60 °C in order to reach the desired viscosity before it is ejected by the nozzle [17]. MJ requires the support material to strengthen primary structures. During 3D printing process, the build platform is gradually moved down by the elevator.

MJ has two main challenges, consisting of droplet formation with a control of droplet deposition. The feedstock materials in this technique should be converted from a continuous liquid into small discrete droplets, which depends on the relationship between devices, materials and processing parameters. Small changes to feedstock materials such as the inclusion of nanoparticles can significantly alter droplet-forming characteristics, resulting in the change in device configuration [18].

Droplet deposition involves droplet flight path, impact and substrate interaction [19]. Generally speaking, build substrate or print head is moving throughout 3D printing process. Hence, it must be taken into account in the calculation of droplet trajectory. Moreover, the location of droplet arrival, size and velocity affect deposition behaviour. Therefore, it should be controlled via nozzle design and operation [20].

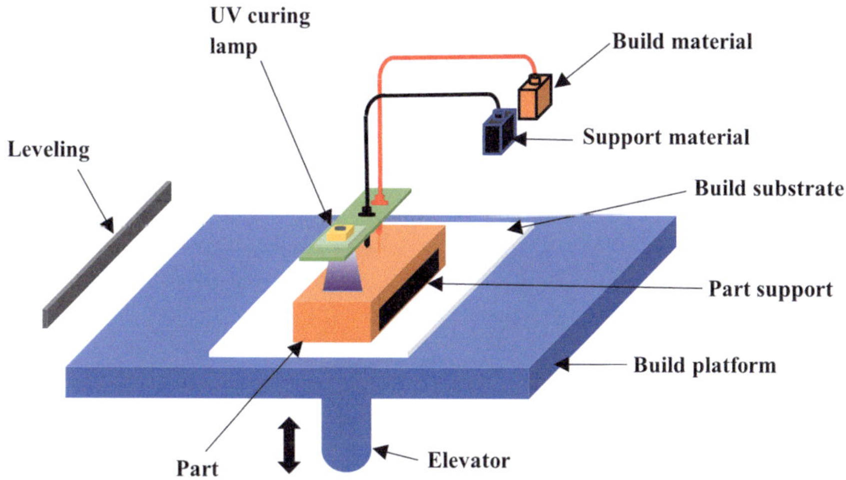

Fig. 1.4 Schematic diagram of material jetting [4]

Three types of MJ printers, consisting of Carmel 700M [21], Production System [22] and Lightspee3D [23], are manufactured by XJet, Desktop Metal and Spee3D enterprises respectively according to Table 1.3. Carmel 700M applies nanoparticle jetting technology to create the components with complex geometries, high accuracy, as well as excellent surface quality, which may reduce the specific requirements for subsequent post-processing. Production System relies on single pass jetting to eject feedstock materials in order to build fully dense and high-quality parts for particular applications. This machine has a minimum layer height of 50 μm and the printing resolution up to 1200 dpi to yield good surface quality and fine features [23]. LightSpee3D depends on supersonic deposition technology to perform quick, cost-effective and scalable production. As reflected by its name, machine nozzle can accelerate air up to three times the speed of sound. The powders are injected and then deposited onto a bed platform at the deposition rate up to 100 g/min [23].

Other printers, like those manufactured by Solidscape company, enable to create the parts with complex geometries and good surface quality such as jewelleries. This company offers S325, S360 and S390 printers that have individual accuracy and

Table 1.3 MJ printers and their manufacturing specifications

Variable	Carmel 700M	Production system	Lightspee3D
Build volume (mm)	500 × 140 × 200	490 × 380 × 260	300 × 300 × 300
Technology	Nanoparticle jetting	Single pass jetting	Supersonic deposition
Printable materials	Nanoparticles (e.g. ceramics, metals)	Metals (e.g. stainless steel, copper, alloy 625)	Metals (e.g. copper, aluminium)
Machine dimension (mm)	3100 × 1850 × 2120	1900 × 5000 × 1900	3130 × 1460 × 2325

resolution up to 25 μm and 5000 dpi respectively. The main differences among such printer types lie in their characteristics of layer thickness and the price. S325, S360 and S390 printers can create layer thicknesses of 25.4, 6.35–38.0 and 25.4–50.8 μm respectively [24].

1.4 Powder Bed Fusion

Similar to DED technique, powder bed fusion (PBF) relies on a thermal source, typically known as an electron beam or laser, in order to effectively fuse feedstock materials in powder form to create an object. In PBF, thermal source selectively melts and sinters powder particles on the specific spots. This 3D printing technique applies a blade recoater or a levelling roller to smooth powder layers and meanwhile controls the distribution of powder fusion in each layer. When a structure layer is formed, the bed platform is moved down, and the levelling roller points the powders to the bed platform [25]. This process is repeated until a part is entirely built, as shown in Fig. 1.5.

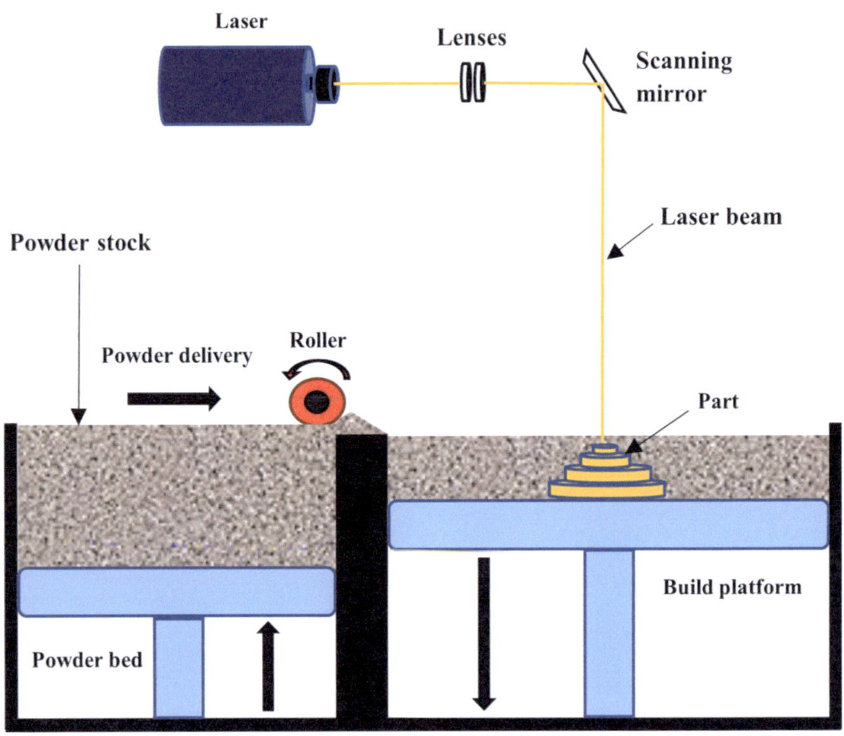

Fig. 1.5 Schematic diagram of powder bed fusion [4]

Currently, PBF can manufacture a part originating from various materials including metals, ceramics and polymers. Polyamide (also known as nylon) is the most commonly used polymer in this technique. This material has different melting points for reliable processibility. Other feedstock materials are derived from different polymers, namely polyaryletherketone (PAEK), polyester (PS), thermoplastic elastomer (TPE) and polypropylene (PP) according to EOS GmbH [26]. Metal materials that can be processed by PBF are aluminium, titanium, cobalt chrome, tool steels, nickel alloys and stainless steel. Meanwhile, ceramics and metal/ceramics have been employed as feedstock materials in PBF for the research purpose [12].

In PBF, when 3D printing process is finished, unbound powders are eliminated using a post-processing step. The powders are applied to support a structure during this process. Cleaning unbound powders is not always easy, mainly for a structure with complex geometry. The main binding mechanism in PBF comprises two processes, namely sintering and melting. Sintering represents partial melting where powder particles melt only at the surface, leading to the porosity on the parts. This porosity can be minimised via a melting process with the incorporation of all particles. Consequently, it enhances tensile and impact strengths of 3D printed parts. The second process of melting is referred to as full melting [27]. In this process, a material region under impinging heat energy melts until its depth surpasses the layer thickness. In addition, the thermal energy of subsequent scans, originating from an electron beam or laser, is sufficient to remelt a portion of a previously hardened structure. As a result, complete melting can yield well-bonded and high-density structures.

Four PBF machines based on selective sintering (SLS) technique are listed in Table 1.4. They are manufactured by four different companies. Kit [28] and Lisa [29] are produced by Sintratec and Sinterit, while Fuse 1 [30] and SnowWhite [31] are manufactured by Formlabs and Sharebot respectively.

Table 1.4 PBF printers and their manufacturing specifications

Variable	Kit	Lisa	Fuse 1	SnowWhite
Build volume (mm)	100 × 100 × 100	110 × 150 × 150	165 × 165 × 320	100 × 100 × 100
Print speed (mm/h)	11	10	10	35
Minimum layer thickness (mm)	0.1–0.15	0.075	0.1	0.05
Laser type	Laser diode	Laser diode	Laser diode	CO_2
Printable material	PA 12, TPE	TPE, PA	PA (nylon)	PA, TPU
Technology	SLS	SLS	SLS	SLS
Machine dimension (mm)	520 × 380 × 600	620 × 400 × 660	677 × 668 × 1059	500 × 500 × 800
Lase power (W)	2.3	2.5	10	14

Among those printers, SnowWhite consumes the highest laser power due to the use of laser CO_2. It also can build the lowest layer thickness with the highest top print speed. Fuse 1 offers the largest build volume, while the smallest build volume is provided by Kit and SnowWhite. According to Table 1.4, Kit, Lisa and Fuse 1 are applicable for the educational purpose, which is opposed to the professional use by SnowWhite. Other PBF printers like ProX DMP 200 Dental, Prodways Promaker P1000, Renishaw RenAM 500E and EOS Formiga P 110 can meet individual requirements for industrial applications.

1.5 Sheet Lamination

Sheet lamination (SL) employs feedstock materials such as plastics, metals and ceramics in sheet form. Silicon carbide (SiC), polyvinyl chloride (PVC), titanium carbide-nickel (TiC-Ni) composites, stainless steel and aluminium represent some typical examples that can be used as feedstock materials in SL. Such a technique consists of two manufacturing processes, namely laminated object manufacturing (LOM) and ultrasonic additive manufacturing (UAM). In LOM process, the lamination of paper material sheets is sliced by laser CO_2 layer by layer where each sheet is one cross-sectional layer of CAD model. The sheet sections not to be part of CAD model must be cut into the cubes via a crosshatch-cutting operation.

UAM is also called ultrasonic consolidation (UC) since it integrates ultrasonic metal welding and computer numerical control (CNC) milling. In UAM, a part is built on a held base plate, which is bolted onto a heated platen. It is generally manufactured layer by layer from the bottom to the top. This platen is commonly heated up to 200 °C [12]. Each layer includes several metal foils being placed side by side, which is then trimmed by using CNC milling.

In UAM process, a rotating sonotrode is moved along the length of a metal foil in a thickness range of 100–150 μm. The foil is held tightly in contact with the prior layer or base plate by implementing a normal force via the rotating sonotrode shown in Fig. 1.6. This sonotrode transversely oscillates in the motion direction with the frequency of 20 kHz at a specific amplitude set by the users [12]. A foil is deposited next to the previous one, which is then repeated until a complete layer is created. A new layer is placed above the previously deposited layer and well bonded according to the similar procedure. In general, one level comprises four foil layers. Once one level is deposited, a CNC milling head constructs the deposited foils to their slice contour. The combination of such additive and subtractive manufacturing processes occurs till the part is fully built.

Not many SL printers are available in the market, which might be caused by their inability to build complex geometries with high accuracy, leading to the limitation of their widespread use. Nonetheless, several companies still continue to manufacture SL printers like EnvisionTEC and Impossible Object to fabricate SLCOM1 [32] and CBAM2 [33] respectively. Meanwhile, manufacturing specifications of both machines are listed in Table 1.5.

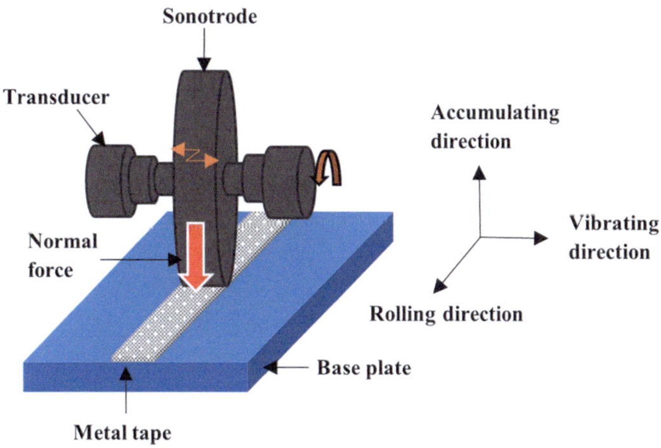

Fig. 1.6 Schematic diagram of ultrasonic additive manufacturing [4]

Table 1.5 SL printers and their manufacturing specifications

Variable	SLCOM1	CBAM2
Build volume (mm)	762 × 610 × 610	305 × 305 × 102
Accuracy (mm)	0.1	0.125
Print speed	508 mm/s	45 in^3/h
Minimum/maximum layer thickness (mm)	0.1 to 1.0	0.05
Printable material	Thermoplastics (PC, PA), continuous fibres	Polyether ether ketone (PEEK), PA12, carbon fibres and glass fibres
Machine dimension (mm)	3863 × 2383 × 3175	3560 × 980 × 2120
Technology	Selective lamination composite object manufacturing (SLCOM)	Composite-based additive manufacturing (CBAM)

SLCOM1 implements a manufacturing technology for selective laminated composite parts that can yield excellent surface quality and mechanical strength, originating from woven fibre composites when pre-impregnated with thermoplastics. This machine also possesses big build volume, good accuracy and high print speed. On the other hand, CBAM-2 enables to produce additively manufactured composite parts with high material performance. Even though CBAM-2 has smaller build volume than SLCOM1, it provides lower layer thickness, which enables this machine to build good parts with fine features.

1.6 Vat Photopolymerisation

Vat photopolymerisation (VP) is capable of processing feedstock materials from radiation-curable resins or photopolymers utilising UV light or laser via polymerisation to build a desired part. A photopolymer can be defined as a special polymer that can convert its characteristics or properties when exposed to visible or UV light. Additionally, polymerisation represents a formation process of polymers derived from basic monomers.

According to the build platform and the laser placement, VP technique is divided into bottom-up and top-down processes. For the bottom-up process, the laser is located on the top of the vat, and the bed platform lowers a part being manufactured after a new selective layer is solidified [34], as shown in Fig. 1.7. Meanwhile, in the top-down process, the laser is placed below the vat, and the bed platform is moved downward until it approaches the vat bottom where the distance between the bed platform and the vat bottom is approximately in size of a thin layer. Laser beam solidifies the thin layer first as the part top layer. The bed platform is then lifted so that the resin can fill a gap between the vat bottom and last solidified layer. Those steps are repeated until a final part is created completely [4].

VP employs photopolymers that have cross-linking capability without melting so that they experience less creep and stress relaxation, which is different from thermoplastics in a branched or linear molecular form that may repeatedly undergo melting and solidification. These photopolymers consist of five elements, namely photoinitiators, reactive diluents, flexibilisers, stabilisers and liquid monomers [12]. A photoinitiator refers to a molecule or a compound that can yield reactive species

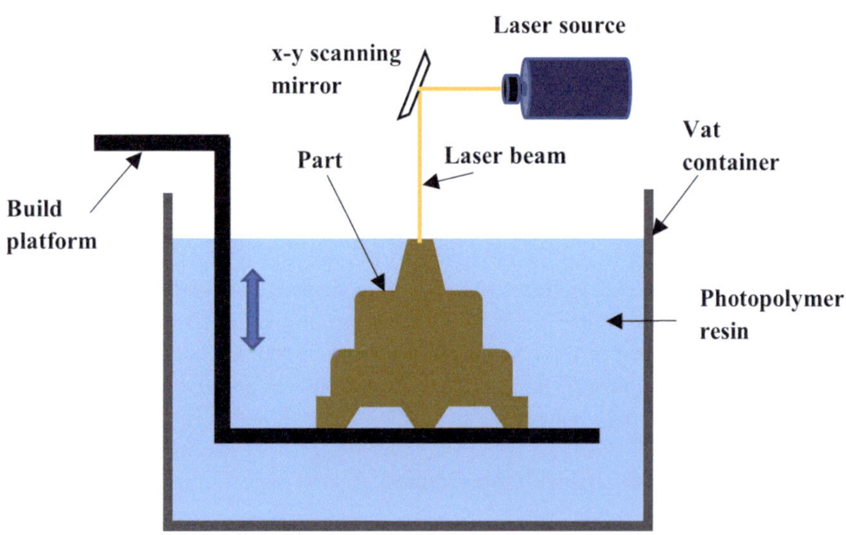

Fig. 1.7 Schematic diagram of vat photopolymerisation using a bottom-up process [4]

Table 1.6 VP printers and their manufacturing specifications

Variable	Photon zero	Shuffle XL lite	Sonic 4K	Inkspire
Build volume (mm)	97 × 54 × 150	192 × 120 × 200	133 × 75 × 200	132 × 74 × 175
Maximum print speed (mm/h)	30	20	200	36
Minimum layer thickness (mm)	0.01	0.01	0.01	0.025
Light source (nm)	405	405	405	405
Printable material	UV resins	UV resins	UV resins	UV resins
Technology	LCD-based SLA	LCD-based SLA	LCD-based SLA	LCD-based SLA

when exposed to the radiation. A reactive diluent is a typical material to reduce a lacquer viscosity, which becomes part of the lacquer during the curing process. A flexibiliser is defined as a substance that can increase the spacing between crystalline polymeric chains to produce flexible and tougher polymers. Meanwhile, chemical additives incorporated into polymers to avoid their degradation are also known as a stabiliser.

Table 1.6 presents four types of VP machines, consisting of Photon Zero [35], Shuffle XL Lite [36], Sonic 4K [36] and Inkspire [37] based on LCD-based stereolithography (SLA) method with their manufacturing specifications. Such machines are produced by three different companies, namely Anycubic, Phrozen and Zortrax. Anycubic and Zortrax manufacture Photon Zero and Inkspire respectively, while Phrozen creates Shuffle XL Lite and Sonic 4K. For those beginners, Photon Zero and Shuffle XL Lite should be sufficient to meet their requirements though Sonic 4K and Inspire with higher printing quality could be applicable to professionals and researchers.

1.7 Fused Deposition Modelling

Material extrusion method discussed in this chapter particularly focuses on fused deposition modelling (FDM). FDM is one of the most popular techniques among all AM methods due to its capability of processing a variety of materials ranging from metals (e.g. copper, stainless steel and Inconel 625) to polymers such as polylactic acid (PLA), acrylonitrile butadiene styrene (ABS) and polyurethane (PU) in the filament form using low-cost manufacturing.

Filament manufacturers such as Markforged [38] and Stratasys [39] supply metallic filaments like 17-4 PH stainless steel, Inconel 625, copper, H13 tool steel, as well as A2 and D2 tool steel. Such companies also provide composite filaments including ABS-CF10 (carbon fibre-filled ABS), Onyx (carbon fibre-filled nylon) and nylon 12/carbon fibre composites. Those filament types are offered to yield 3D printed parts with desired properties required by FDM printer users.

Main FDM parts comprise four elements, which are known as feed pinch rollers, a heating chamber or a liquefier, a nozzle and a bed platform, as illustrated in Fig. 1.8a. First of all, the filament is fed through a pinch-roller feed system, placed into the heating chamber, extruded via a nozzle tip and finally deposited layer by layer onto a bed platform in order to build a part. Figure 1.8b displays an Axiom20 FDM printer supplied by Airwolf 3D printer company (Las Vegas, USA) [40], which is further employed in this study.

Four types of FDM printers including Ender-3 Pro [41], DIY [42], Finder 2.0 [43] and X-Pro [44] with their individual manufacturing specifications are listed in Table 1.7. Ender-3 Pro and DIY are supplied by Creality and FLSUN. Meanwhile, Finder 2.0 and X-Pro are manufactured by Flashforge and Qidi Tech companies respectively.

Ender-3 Pro has a compact design, high-quality printing, easy assemble, upgradable machine, as well as a tight filament path to enhance the compatibility with the filaments. Furthermore, it provides the highest printing speed with the ability to process more materials when compared with Finder and DIY. Nevertheless, this printer consumes more energy relative to the others, but the consumers are not satisfied with the magnetic bed of Ender-3 Pro. In addition, this printer also has a typical issue with brittle filaments and its print bed levelling.

DIY is capable of fabricating a tall model because of its unique construction and quiet-printing performance via a delta printing technology. Such a technology enables to separate stepper motor to drive each articulating arm. Besides, its print head is small and light, resulting in the occurrence of low noise during printing processes. However, DIY only possesses cramped build space that does not suit flexible filaments

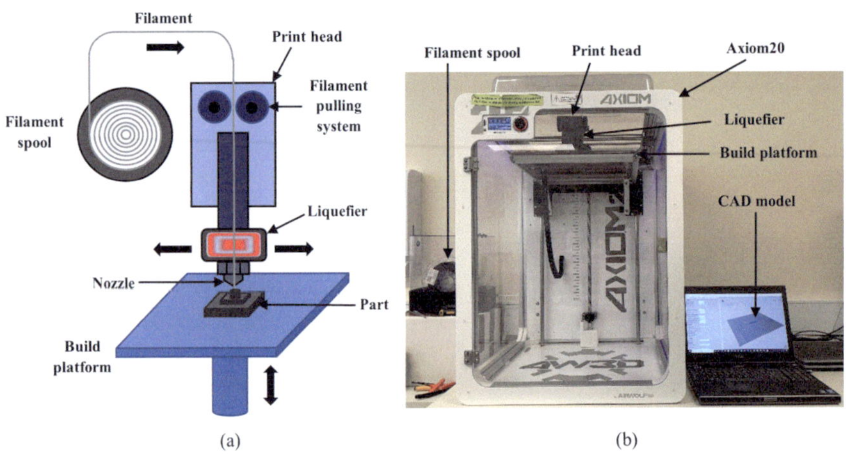

Fig. 1.8 **a** Schematic diagram of directed FDM [4] and **b** FDM printer-Axiom 20

Table 1.7 Specification of four commonly used FDM printers

Variable	Ender-3 Pro	DIY	Finder 2.0	X-Pro
Build volume (mm)	220 × 220 × 250	Ø180 × 300	140 × 140 × 140	230 × 150 × 150
Maximum printing speed (mm/s)	180	150	150	150
Layer thickness (mm)	0.1–0.4	0.05–0.4	0.1–0.5	0.05–0.4
Print resolution (mm)	0.1	0.1	0.1	0.1
Filament diameter (mm)	1.75	1.75	1.75	1.75
Nozzle diameter (mm)	0.4	0.4	0.4	0.4
Maximum nozzle temperature (°C)	255	250	240	250
Maximum bed temperature (°C)	110	100	Not available	250
Printable material	PLA, ABS, wood, TPU and carbon fibres	PLA and ABS	PLA	ABS, TPA, TPU, PETG and HIPS
Machine dimension (mm)	440 × 440 × 465	300 × 400 × 680	420 × 420 × 420	580 × 536 × 470
Weight (kg)	6.9	6.8	11	18
Power consumption (W)	270	200	60	100

with the assembling difficulty. Such drawbacks hinder its widespread use, as opposed to Cartesian 3D printers like Finder 2.0 and Ender-3 Pro.

Finder 2.0 offers quiet printing, low energy consumption, clear design, easy to use, touchscreen and good visibility of print. Nonetheless, its build volume and filament cartridge are too small. Moreover, the print bed is not heated, leading to detrimental warping effect. Moreover, this printer is limited to process PLA and polyethylene terephthalate glycol (PETG).

X-Pro provides two nozzles, allowing such printers to build a final part using two types of filaments. It also can handle various materials like PETG, ABS, thermoplastic polyurethane (TPU), high impact polystyrene (HIPS) and TPA with high printing accuracy. Meanwhile, other printers such as Stratasys F120, Markforged Mark 2 and Ultimaker S5 are categorised as high-grade FDM machines, enabling to handle various materials and build a part with high accuracy, as well as require minimum maintenance. Such printers may meet specific demands for professional, business and industrial purposes.

1.8 Summary

Seven methods of additive manufacturing, consisting of binder jetting, directed energy deposition, material jetting, powder bed fusion, sheet lamination, vat photopolymerisation and material extrusion, are discussed in this chapter. Directed energy deposition, powder bed fusion and sheet lamination employ electron or laser beam as the heat source in order to create a ceramic or metal part. On the other hand, UV light is used by material jetting and vat photopolymerisation to harden photopolymer resin for part manufacturing. Binder jetting utilises ceramic and metal powders along with the liquid binder to build a part. Sheet lamination combines subtractive manufacturing, laminated object manufacturing, as well as ultrasonic additive manufacturing to fabricate a desired part. In particular, fused deposition modelling relies on the material extrusion process and filaments as feedstock materials to build final components in a layer-by-layer manner.

References

1. ISO/ASTM 52900: Additive manufacturing-General principles-Fundamentals and Vocabulary (2021)
2. Attaran, M.: The rise of 3-D printing: the advantages of additive manufacturing over traditional manufacturing. Bus. Horiz. **60**, 677–688 (2017)
3. ISO/ASTM 52900: Additive Manufacturing-General Principles-Terminology. International Organization Standardization (2015)
4. Nugroho, W., Dong, Y., Pramanik, A.: Chapter 4—3D printing composite materials: a comprehensive review. In: Low, I.-M., Dong, Y. (eds.) Composite Materials: Manufacturing, Properties and Applications, pp. 65–115. Elsevier B.V (2021)
5. Sachs, E., Cima, M., Cornie, J.: Three-dimensional printing: rapid tooling and prototypes directly from a CAD model. CIRP Ann. **39**, 201–204 (1990)
6. Zenou, M., & Grainger, L. (2018). Additive manufacturing of metallic materials. In: Zhang, J., Jung, Y.-G. (eds.) Additive Manufacturing: Materials, Processes, Quantifications and Applications. Butterworth-Heinemann
7. Kumar, A., Wang, J., Bai, Y., Huxtable, S., Williams, C.: Impacts of process-induced porosity on material properties of copper made by binder jetting additive manufacturing. Mater. Des. **182**, 1–12 (2019)
8. Voxeljet: Industrial 3D Printing Systems: Efficient and Economical. https://www.voxeljet.com/. Accessed 24 Aug 2020
9. ExOne: Innovent^{+}. https://www.exone.com/en-US/3D-printing-systems/metal-3d-printers/Innovent. Accessed 24 Aug 2020
10. D. Metal: DM P2500. https://digitalmetal.tech/printer-line/dm-p2500/. Accessed 24 Aug 2020
11. ExOne: 3D Printing Technology. https://www.exone.com/en-US. Accessed 24 Aug 2020
12. Gibson, I., Rosen, D., Stucker, B.: Additive Manufacturing Technologies—3D Printing, Rapid Prototyping, and Direct Digital Manufacturing. Springer Science+Business Media, New York (2015)
13. Additec: μPrinter. https://www.additec.net/%CE%BCprinter_ Accessed 24 Aug 2020
14. Formalloy: Machines. https://www.formalloy.com/machines. Accessed 24 Aug 2020
15. Optomec: LENS Systems-3D Printed Metals. https://optomec.com/3d-printed-metals/lens-printers/. Accessed 24 Aug 2020

16. Tay, B., Evans, J., Edirisinghe, M.: Solid freeform fabrication of ceramics. Int. Mater. Rev. **48**, 341–370 (2003)
17. Hutchings, I., Martin, G.: Inkjet Technology for Digital Fabrication, 1st edn. Wiley, Chichester (2012)
18. Furbank, R., Morris, J.: An experimental study of particle effects on drop formation. Phys. Fluids **16**, 1777–1790 (2004)
19. Bechtel, S., Bogy, D., Talke, F.: Impact of a liquid drop against a flat surface. IBM J. Res. Dev. **25**, 963–971 (1981)
20. Zhou, W., Loney, D., Fedorov, A., Degertekin, L., Rosen, D.: What controls dynamics of droplet shape evolution upon impingement on a solid surface? AIChE J. **59**, 3071–3082 (2013)
21. XJet: XJet Carmel 700M Cost-efficient metal AM system. https://www.xjet3d.com/products/metal-systems/. Accessed 24 Aug 2020
22. D. Metal: Production System™. https://www.desktopmetal.com/products/production. Accessed 24 Aug 2020
23. Spee3D: LightSPEE3D. https://www.spee3d.com/product/lightspee3d/. Accessed 24 Aug 2020
24. Solidscape: 3D Printers. https://www.solidscape.com/3d-printers/. Accessed 24 Aug 2020
25. Vyas, C., Poologasundarampillai, G., Hoyland, J., Bartolo, P.: 3D printing of biocomposites for osteochondral tissue engineering. In: Ambrosio, L. (ed.) Biomedical Composites, 2nd edn., pp. 261–302. Woodhead Publishing, Cambridge (2017)
26. EOS: EOS 3D Printing Materials—Plastics for Additive Manufacturing SLS Materials and Processes. EOS. https://www.eos.info/en/additive-manufacturing/3d-printing-plastic/sls-polymer-materials. Accessed 15 Oct 2021
27. Michaels, S., Sachs, E., Cima, M.: Metal Parts Generation by Three-Dimensional Printing. Paper presented at The Fourth International Conferences on Rapid Prototyping, Dayton, 25, 1993 (1993)
28. Sintratec: Sintratec Kit. https://sintratec.com/product/sintratec-kit/. Accessed 24 Aug 2020
29. Sinterit: Meet Sinterit Lisa (2020). https://www.sinterit.com/sinterit-lisa/. Accessed 24 Aug 2020
30. Formlabs: Meet the Fuse 1. https://formlabs.com/asia/3d-printers/fuse-1/. Accessed 24 Aug 2020
31. Sharebot: Sharebot SnowWhite. https://www.sharebot.it/en/sharebot-snowwhite-3d-printer/. Accessed 24 Aug 2020
32. Envisiontec: SLCOM1. https://envisiontec.com/de/3d-printers/slcom-1-de/slcom-1/. Accessed 24 Aug 2020
33. Objects, I.: The CBAM-2 Printer. https://www.impossible-objects.com/cbam-printer/. Accessed 24 Aug 2020
34. Guelcher, S.: Biodegradable polyurethanes: synthesis and applications in regenerative medicine. Tissue Eng. Part B **14**, 3–17 (2008)
35. A.D. Printing: Photon Zero. https://www.anycubic.com/products/anycubic-photon-zero. Accessed 24 Aug 2020
36. Phrozen: Phrozen. https://www.phrozen3dp.com/?locale=en. Accessed 24 Aug 2020
37. Zortrax: Zortrax Inkspire Made for Speed and Surgical Precision. https://zortrax.com/3d-printers/inkspire. Accessed 24 Aug 2020
38. Markforged: 3D Printing Materials. Markforged. https://markforged.com/materials. Accessed 16 July 2021
39. Stratasys: Our Materials. Stratasys (2021). https://www.stratasys.com/materials/search. Accessed 16 July 2021
40. Airwolf3D: 3D Printers. Airwolf 3D Printers. https://airwolf3d.com/product-category/3d-printers/. Accessed 16 July 2021
41. Creality: Ender 3 Pro 3D Printer. https://www.creality3dofficial.com/products/creality-ender-3-pro-3d-printer?gclid=EAIaIQobChMI2Iftjc2w6wIVSCQrCh0_EANlEAAYASAAEgJMnfD_BwE. Accessed 24 Aug 2020

42. FLSUN: FLSUN DIY 3D Printer Kit of Kossel Delta with Heated Bed, 1KG PLA Filament and LCD Display. http://www.3dflsun.com/goods.php?id=67. Accessed 24 Aug 2020
43. F.D. Printer: Flashforge Finder 3D Printer with Cloud, Wi-Fi, USB for Education. https://www.flashforgeshop.com/product/10003?gclid=EAIaIQobChMIz_Ctm8yw6wIVhw4rCh3nSwPtEAAYASAAEgIOFPD_BwE. Accessed 24 Aug 2020
44. Q. Technology: X-PRo. https://www.qd3dprinter.com/products/x-pro/. Accessed 24 Aug 2020

Chapter 2
Fundamental Steps in Additive Manufacturing

Abstract Chapter 2 covers five fundamental steps in additive manufacturing. First of all, 3D models are created using CAD software. Such models are then converted into standard tessellation language (STL) format so that they can be processed by 3D printers. The next step is to set up printing parameters of 3D printers to meet stringent requirements of end-users. Once 3D printing set up is completed, the manufacturing process can be started. Removal and clean-up processes are conducted after 3D printed part is entirely built. Some printed parts rely on the post-processing step in order to further improve their surface finish and mechanical strength.

Keyword Additive manufacturing · 3D printer · 3D CAD model · Standard tessellation language (STL) · Post-processing

In general, additive manufacturing (AM) like fused deposition modelling (FDM) comprises several steps and iterations throughout printing and product improvement processes. These steps depend on the size, complexity of geometry, surface finish and material properties of desired parts. For small, rough and simple parts, basic manufacturing stages should be sufficient to meet the requirements of end-users. Nonetheless, for a large and complex structure with excellent mechanical strength and fine surface finish, an additional step of post-processing should be generally taken [1]. Overall, AM process is divided into six steps in the following Sects. 2.1–2.5.

2.1 Creation of 3D Models

3D model is designed via CAD software that allows the designers to create, modify and optimise it. Furthermore, such software enables the designers to increase design quality, enhance their productivity and establish a manufacturing database. It is noted that all geometries along with the shape of a model must be completely described in CAD software. This signifies that CAD data should be "watertight" without gaps, voids or holes, and overlap in model geometries [2], as illustrated in Fig. 2.1a. CAD

W. T. Nugroho et al., *Thermoplastic Polyurethane (TPU)/Halloysite Nanotube (HNT) Nanocomposites*, Engineering Materials, https://doi.org/10.1007/978-981-95-0541-8_2

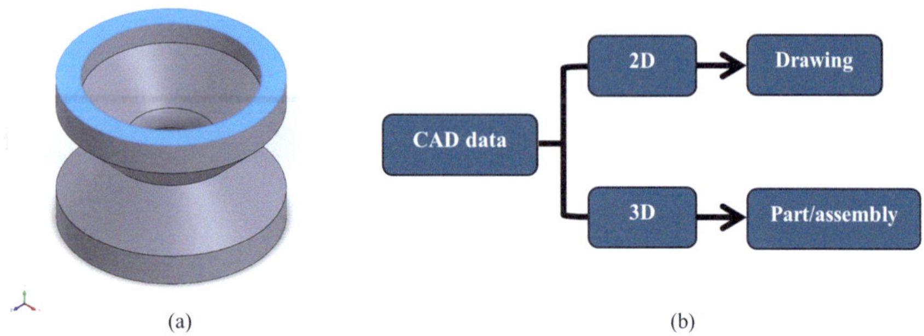

Fig. 2.1 **a** Typical CAD model and **b** CAD data use

data typically contains 2D or 3D design, which are employed to set up models or architecture plans shown in Fig. 2.1b. In relation to this, 2D CAD data refers to specific drawings, while 3D CAD data is well known as parts or assemblies.

Before creating CAD models, CAD system architecture should be known in advance in order to provide comprehensive insight for the modelling process using CAD software. This system consists of three main elements, namely graphic user interface (GUI), operating system and CAD data (database) [3]. GUI is constructed using non-uniform rational basis (NURB) geometry/boundary representation, known as B-rep, with the aid of a geometric modelling kernel. The kernel is a principal factor in bolstering underline computing and also modelling ability for solid objects. Besides, it can export solid models from 3D geometry. Commercially available solid modelling systems are established based on a geometric modelling kernel, also known as a modelling engine or a geometry library. It refers to a library of core mathematical functions that builds and stores 3D solid objects in order to respond to the user's commands. Such command inputs are then processed by kernel library via the interface of end-users, store the data output, and submit it to the graphics package for display [4]. Nowadays, CAD software is capable of supporting various operating systems such as Windows, Mac and Linux. In addition, this software can handle numerous tasks with minimal hardware requirements, good accessibility and easy use.

CAD software manufacturers provide several software options to design a model for additive manufacturing. Autodesk promotes Netfabb for the purpose of design and simulation instead of design-only AutoCAD [5]. Dassault Systèmes develops CATIA to construct a sophisticated model [6]. Siemens, Parametric Technology Corporation (PTC) and Robert McNeel & Associates (TLM, Inc.) offer Solid Edge [7], Onshape [8] and Rhino [9] for designing and simulating 3D models respectively, as listed in Table 2.1. All software has particular advantages and drawbacks according to their characteristics. In particular, CATIA, Solid Edge and Netfabb yield excellent modelling and simulation. However, they require a powerful computer setup with high cost. On the other hand, Onshape and Rhino can be classified into low-cost CAD software despite the provision of minimal facilities in modelling and simulation.

Table 2.1 Holistic comparison of CAD software

CAD software	Manufacturer	Application	Price (US$) per year	Advantage	Disadvantage	Refs.
Netfabb	Autodesk	Additive manufacturing, design and simulation	5300	• Powerful workflow automation for streamlining 3D print preparation • Intuitive tools for repairing, orienting, arranging and preparing 3D files for printing • Agile latticing and hollowing tools and parametric supports for light weight	• Requiring powerful computer setup • Requiring time and effort to become proficient in using the software effectively due to the extensive features and complex tool sets of Netfabb • Stability and performance issues	[5]

(continued)

Table 2.1 (continued)

CAD software	Manufacturer	Application	Price (US$) per year	Advantage	Disadvantage	Refs.
CATIA	Dassault systèmes	Fusing modelling and simulation	7560	• Capable of designing, simulating and analysing complex engineering systems, such as aircraft, ships, cars, industrial machinery and consumer goods • Allowing to streamline and accelerate design, analysis and data management • Capable of optimising the use of computer resources for engineers to retrieve large assemblies quickly without experiencing delays or slowdowns	• Less user-friendly interface • Difficulty in product modelling • Difficulty in mouse operations • Heavy software with a long wait time to load files and making changes • Difficult in updating an existing design	[6]
Solid edge	Siemens	For 3D design, simulation, manufacturing, data management, and cloud collaboration	5772	• Easy to use • Rapid and smooth data processing • Capable of running multiple applications	• Bit messy interface • Slow booting process • Unrecoverable errors	[7]

(continued)

Table 2.1 (continued)

CAD software	Manufacturer	Application	Price (US$) per year	Advantage	Disadvantage	Refs.
Onshape	Parametric Technology Corporation (PTC)	For product development platform with integrated CAD, data management and analytics	2500	• Excellent user flow, as well as very high application efficiency and stability capable of designing and producing complex 3D printed objects to fit together perfectly	• A lack of offline caching, which means models need to be constantly reloaded when experiencing poor connectivity issues • A lack of private projects with free accounts	[8]
Rhino	Robert McNeel & Associates (TLM, Inc.)	3D modelling and design application	995	• Easy to use • Good output to present great tools and customisation interoperability with other software and easy-to-share files	• Having the lack and difference in some features and graphical user interface (GUI) from the Windows version that may prove confusing when switching between both operating systems • Very hard to figure out the reason for a procedure failure to end up spending a lot of time in trying various other paths to reach the desired end • Not intuitive user interface	[9]

Meanwhile, 3D printer providers like 3D Systems [10], Markforged [11] and Airwolf [12] have developed engineering software suitable for their printers. 3D Systems offers 3DXpert for AM support, while Markforged provides Eiger as a standard 3D printing and Blacksmith that may be employed for artificial intelligence used in adaptive manufacturing software. Apex, a 3D printing software from Airwolf, is exclusively designed for Airwolf 3D printers, which are well known as Evo and Axiom series [12].

2.2 Conversion of 3D Models

Generally speaking, a 3D model file in the format of CAD data must be converted into STL format [2]. This STL file acts as a connector between the CAD system and 3D printers. In other words, AM process cannot read specific data acquired from the CAD system without the aid of STL file. Meanwhile, the conversion of a CAD model into STL format typically induces a gap between model surface and triangle face, which is known as facet deviation shown in Fig. 2.2. The accuracy level of STL file is represented by the facet size. Smaller facet size reduces the deviation leading to better model accuracy. Figure 2.3 reveals that left circle (A) with bigger triangle size generates larger facet deviation than the right one (B). Accordingly, the main purpose for the conversion is to reduce facet deviation. Facet deviation for AM should be typically in range of 0.025–0.05 mm [2].

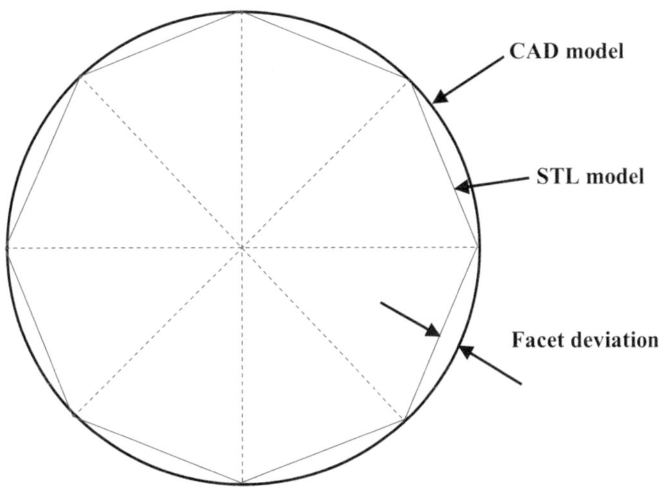

Fig. 2.2 STL format conversion [1]

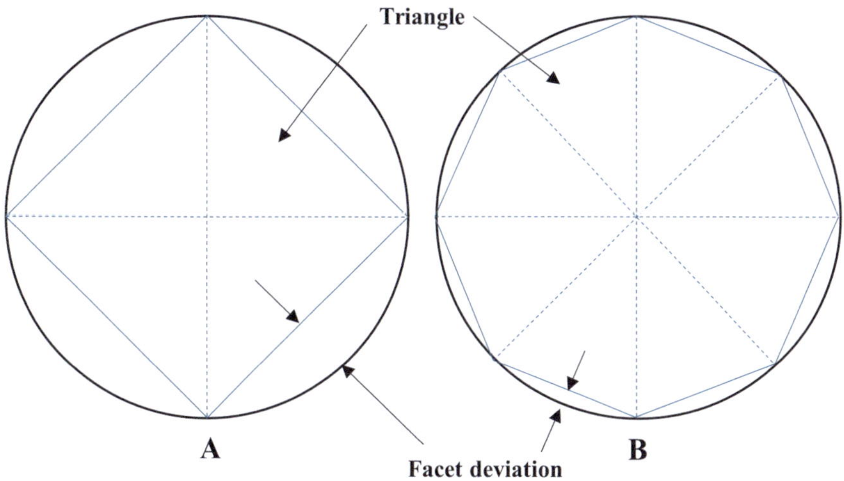

Fig. 2.3 Facet deviation induced by different triangle sizes

2.3 3D Printer Configuration

The printing parameters of 3D printer need to be set up properly before any manufacturing process. This section discusses three printing parameters, namely layer height, nozzle speed and extrusion rate. The setup of layer height aims to adjust it by setting up the z-axis parameter in G-code. In general, the steps to adjust G-code are displayed in Fig. 2.4. Basically, if a particular section such as layer height is changed, this part should be copied and then pasted into Notepad ++ . In this software, layer height data are usually adjusted and copied. The modified data are generally used to replace initial information in G-code. The better dimensional accuracy can be often achieved by the thinner layer. Nevertheless, it requires longer fabricating time leading to high manufacturing cost. The nozzle speed influences printing efficiency and effect [13].

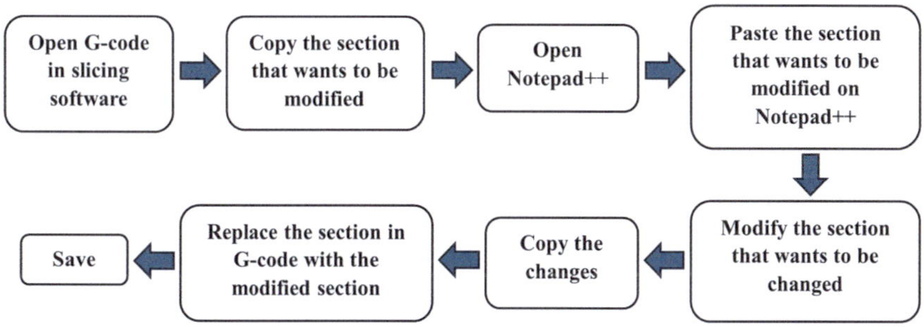

Fig. 2.4 Flowchart of G-code modification

Theoretically, the optimum printing speed can be calculated through critical nozzle speed by using Eq. (2.1) as follows

$$v_n = \frac{4Q}{\pi D_N^2}, \tag{2.1}$$

where v_n, Q and D_N denote critical nozzle speed (mm/s), material extrusion rate (cm^3/s) and nozzle diameter (mm). When nozzle speed is higher than v_n, it creates printed filaments with smaller diameter than nozzle diameter [14]. Consequently, it yields a printed part with smaller total height and low precision. On the other hand, if nozzle speed is lower than v_n, it increases layer height, causing the nozzle tip to hit a printed part. The optimum result can be obtained when nozzle speed is equal to v_n where printed filament size is similar to nozzle diameter.

Extrusion rate represents material extrusion speed per second, which is connected with step motor rate used for controlling an extruder piston. A high extrusion rate in range of 0.004–0.005 cm^3/s results in a filament diameter larger than D_N [15]. It induces the feedstock material to be distributed to an entire build platform before its setting. On the contrary, a low extrusion rate (e.g. 0.002 cm^3/s) can reduce the extrusion pressure, thus causing inconsistently extruded filaments [16]. Hence, an optimal extrusion rate is crucial to produce a proper filament diameter equal to D_N. 3D printer configuration with the used parameters such as layer height, fill density, print speed, printing temperature, bed temperature along with x, y and z axes is illustrated in Fig. 2.5.

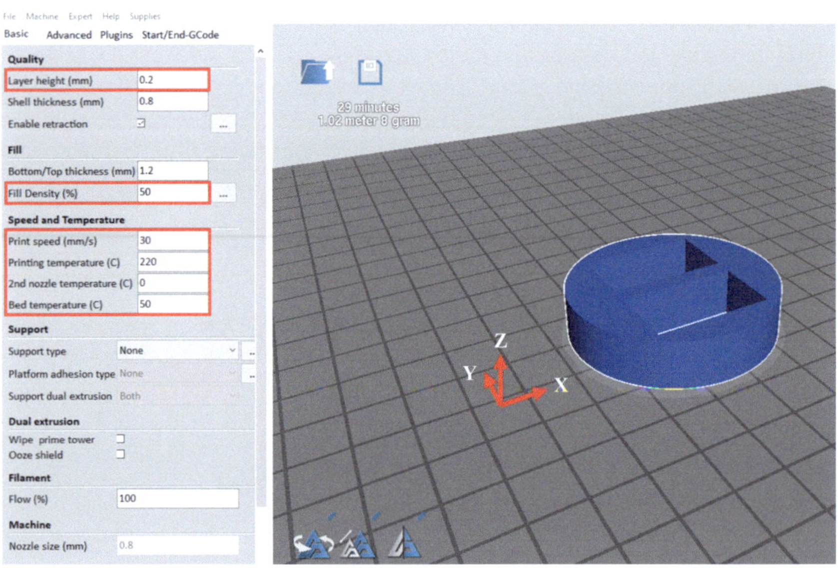

Fig. 2.5 3D printer configuration in x, y and z axes

2.4 Build Process of Feedstock Materials

In the preparation step, especially for FDM process, feedstock materials in the filament form are set up at the spool holder. Setting 3D printing parameters such as printing temperature, printing speed, layer height, infill density, etc., should be conducted before starting printing. Preheating, which is known as heating the printer until reaching operating temperature, is typically required to facilitate a filament-feeding process. Afterwards, it continues with an initial extruding process where the filaments are ejected through a nozzle tip. This step is intended to ensure that extruded filaments can meet stringent requirements of end-users. After all preparation steps shown in Fig. 2.6 are conducted, a build process can be commenced.

Human labour is not necessary in the build process for an automated process. Nonetheless, regular monitoring by a human is still required to ensure that the building process can run well without any errors, such as running out of materials and software glitches, which is considered as one of most common 3D printing merits. Human labour is typically necessary to prepare feedstock materials for a 3D printer by setting up filaments on the spool holder and then removing printed parts upon the

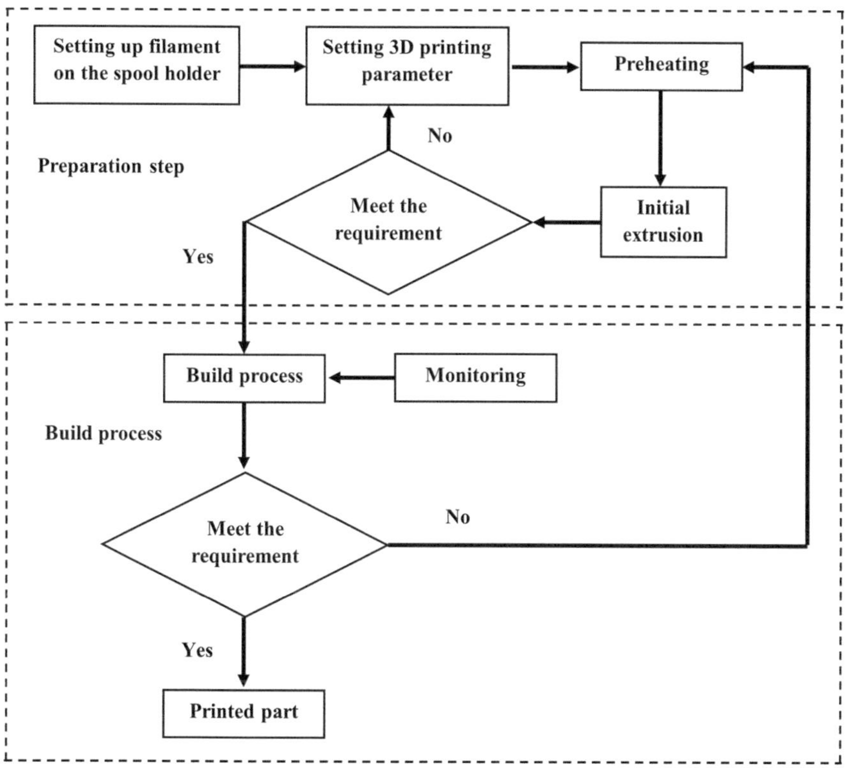

Fig. 2.6 Flowchart of preparation and build process in FDM

completion of the manufacturing process. Once the monitoring result indicates that the build process meets the desired requirements, this process can continue until a part is fully constructed.

As mentioned earlier in Chapter 1, 3D printing methods can be divided into seven categories, namely binder jetting (BJ), directed energy deposition (DED), material jetting (MJ), powder bed fusion (PBF), sheet lamination (SL), vat photopolymerisation (VP) and fused deposition modelling (FDM) as the representative for material extrusion (ME). BJ, DED, PDF, SL and FDM use powder particles, sheets and filaments, as opposed to MJ and VP employing liquid materials. The feedstock materials selected for such 3D printing methods are listed in Table 2.2.

Table 2.2 Selected materials in 3D printing

3D printing method	Main material	Form
Binder jetting (BJ)	Metals (titanium, copper and steel) Ceramics (carbon, carbide, alumina, silicon and sands)	Powder particles and liquid binders
Directed energy deposition (DED)	Tungsten carbide, titanium carbide, continuous carbon fibre-reinforced thermoplastics	Powder particles and wires
Material jetting (MJ)	Photopolymer resins	Ink (liquid)
Powder bed fusion (PBF)	Metals (aluminium, titanium, cobalt, chrome, tool steels Ceramic (alumina) Polymer (polyaryletherketone, polyester, thermoset elastomer and polypropylene)	Powder particles
Sheet lamination (SL)	Silicon carbide, polyvinyl chloride, titanium carbide-nickel composites	Sheets
Vat photopolymerisation (VP)	Photopolymer resins	Liquids
Fused deposition modelling (FDM)	Metals (copper, stainless steel and Inconel 625) Polymers (polylactic acid, acrylonitrile butadiene styrene and polyurethane)	Filaments

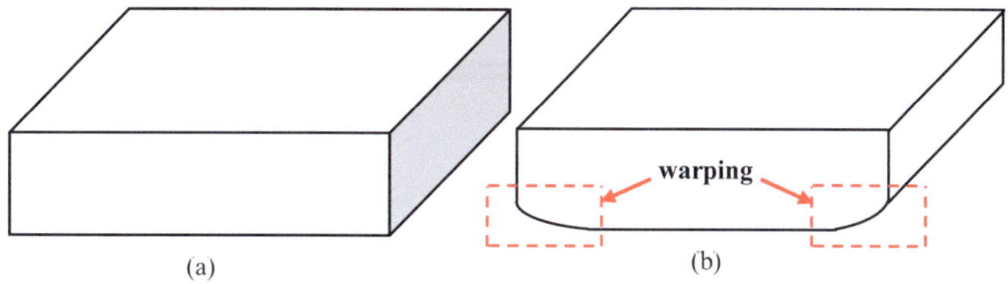

Fig. 2.7 3D printed part: **a** without and **b** with warpage effect

2.5 Removal, Clean-Up and Post-Processing

2.5.1 Removal and Clean-Up

Once the entire structure is printed, removal and clean-up are performed. In particular, a simple structure in FDM process typically can be easily moved from the build platform when it does not apply an adhesive material such as Wolfbite [17]. The addition of adhesive materials aims to enhance the clinginess between extruded filaments and the build platform. This addition is usually required for a long-time printing process over 3 h and the build platform of FDM printers made from glass materials. In such a condition, extruded filaments might undergo the warpage shown in Fig. 2.7, which affects dimensional accuracy of 3D printed parts, particularly for their thickness. The clean-up of 3D printed part and build platform is crucial in order to yield optimum results and facilitate next printing process.

2.5.2 Post-Processing

It is evidently shown that additive manufacturing, also known as 3D printing, has many advantages, such as the fabrication of a part with complex geometry and many open cavities like lattice structures. However, 3D printing technology also exhibits some disadvantages. One of the most important demerits of this kind of technology is to produce parts with inferior surface quality. The parts made by various 3D printing processes produce different values of surface quality. Accordingly, 3D printing alone cannot fabricate parts that simultaneously meet the requirements for mechanical properties and surface quality.

The parts manufactured by 3D printing are unable to be employed as end products due to attached support structures, low surface quality and dimensional accuracy, as well as staircase effects [18]. As such, post-processing steps are required to improve surface quality and mechanical properties of 3D printed parts for the use-readiness. Sometimes, a single post-processing method is not sufficient to increase the quality of 3D printed parts. Therefore, combining more than one method is conducted to obtain required quality of such parts. In relation to this, this book divides post-processing processes into four categories, namely mechanical, chemical and electrochemical, thermal and hybrid methods.

2.5.2.1 Mechanical Post-Processing Method

In mechanical post-processing methods, both conventional and non-conventional manufacturing processes have been implemented owing to their rapid and mature nature, as well as good accessibility. For conventional mechanical post-processing, CNC milling [19], vibration-assisted machining [20], ultraprecision machining [21] and grinding have been utilised by the researchers to finish 3D printed parts. Meanwhile, for non-conventional mechanical post-processing, microblasting with high air pressure [22], magnetic abrasive finishing (MAF) [23], ultrasonic assisted abrasive polishing [24] and abrasive flow machining (AFM) [25] are often employed, as depicted in Fig. 2.8a, to eliminate loose and unmelted particles and balling effects. Besides, this hybrid process is beneficial to finish the outer appearance of the parts.

2.5.2.2 Chemical and Electrochemical Methods

The chemical post-processing method is recognised to be affordable and straightforward to use [26]. This method is typically applied in the biomedical industry to clean and polish 3D printed part surfaces. It is well known that certain chemical solutions can dissolve plastic parts, as evidenced by the parts made from acrylonitrile butadiene styrene (ABS) via FDM. In a chemical method, tools and part surfaces do not touch each other. As such, it can enhance dimensional and geometric stability. Nevertheless, it has been observed that various chemical post-processing methods undermine the tensile strength of 3D printed parts and also deteriorate final part geometry [27].

An electrochemical polishing can be defined as a dissolution process, which is conducted in an electrolytic cell. The electrochemical reaction occurs between an electrode functioning as a cathode and 3D printed part (workpiece) as an anode [see Fig. 2.8b]. The potential difference between both cathode and anode should be maintained in a DC range of 2–20 V [28]. During this process, the amount of workpiece material is eliminated in the form of microparticles. It is noteworthy that there is no physical contact between tools and workpieces throughout the process. Therefore, it makes electrochemical polishing method suitable for processing fragile, complex and hardened materials performed by mechanical methods at the post-processing step such as blasting. Both electrolyte and electrode are selected based on the materials used for 3D printing processes [29].

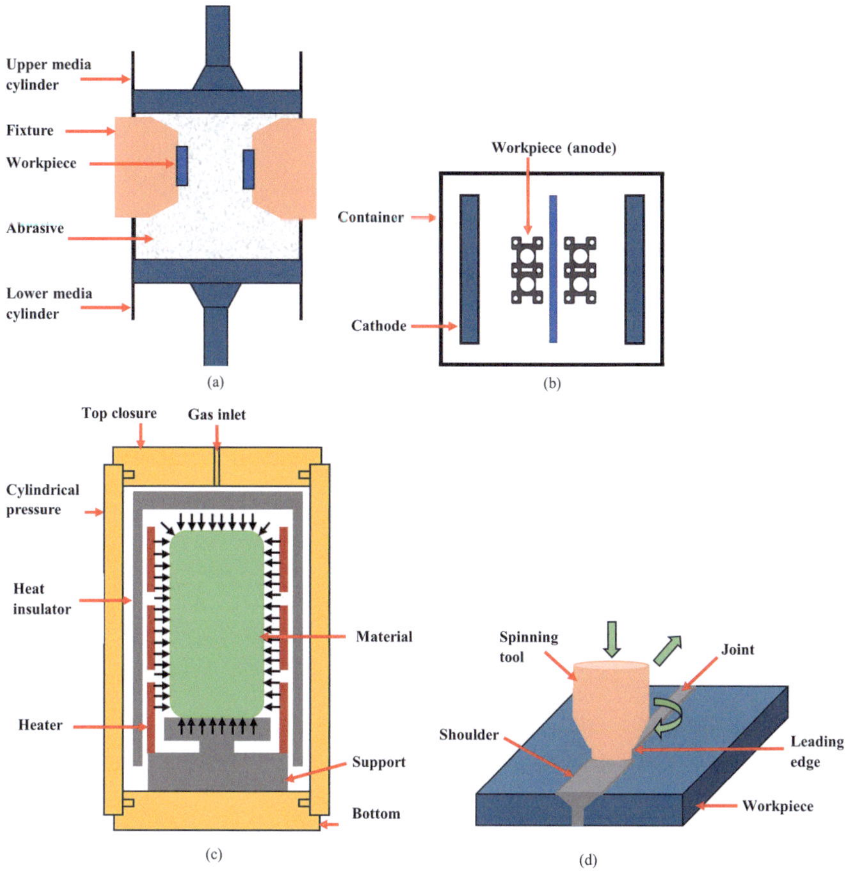

Fig. 2.8 Schematic diagram of post-processing methods: **a** Abrasive flow machining (AFM) process, **b** electrochemical post-processing method, **c** hot isostatic pressing (HIP) process and **d** friction stir welding

2.5.2.3 Thermal Post-Processing Method

Thermal post-processing is capable of minimising cracking, residual stresses and microstructural heterogeneity in 3D printed parts. This method involves three typical processes, consisting of solution heat treatment (SHT), hot isostatic pressing (HIP) and T6 heat treatment (T6 HT), which can significantly improve the part quality made from AlSi10Mg. In HIP the part is evenly pressed from all orientations at a high temperature according to Fig. 2.8c. As such, it yields higher density, better uniformity and excellent performance. In this process, the temperature level approaches up to 1000–2000 °C, while working pressure can reach as high as 200 MPa via an inert gas filled in a container. HIP process allows for the treatment or the removal of pores and intrinsic flaws in 3D printed parts via PBF [30].

2.5.2.4 Hybrid Post-Processing Method

In the hybrid post-processing method, two different instruments, methods, or energy systems are integrated into a single system. For instance, CNC milling and AM have been combined to gain the direct benefit of both processes. Based on this method, the requirement of material deposition and disposal are handled in a single station, which is known as hybrid rapid prototyping (RP) [31]. In the secondary processes used with AM, we only discuss about three popular hybrid AM processes, namely friction stir AM, hybrid additive manufacturing (HAM) by remelting and HAM by ablation.

Friction stir AM, represented by friction stir welding (FSE), can be categorised as a sequential HAM technique, as illustrated in Fig. 2.8d. FSW is implemented to permanently join two surfaces. In this method, a non-consumable spinning tool derived from refractory materials is combined with the workpiece [29]. Due to the pressing process, heat is generated and a large quantity of permanent deformation takes place. It usually induces recrystallisation and substantial grain refinement in metals, which further accelerates fatigue and mechanical performance. Other materials, such as polymers and composites, can also be processed in a similar manner.

HAM by remelting involves the use of an energy source using a laser or an electron beam to remelt previously fused materials. No material is taken out in this process, but it is changed instead [31]. Remelting increases the prior set material to the melting point resulting in better bond layers together. During 3D printing process, pores are created, and the melted material then fills them to yield parts with a higher density over 99%. Remelting can be carried out after a layer or a few layers are formed. A remelted area of the material undergoes the changes in its physical, mechanical and chemical properties. These settings depend on the number of factors, namely power, scanning pattern and laser speed.

HAM by ablation removes the top layer of accumulated materials using an electron beam or a laser [29]. An integrated non-contact technique known as ablation between constructed layers increases the performance and quality of the parts. By removing materials, this technique can produce smooth and precise surfaces.

2.6 Summary

Additive manufacturing commonly involves five main steps, which is initiated with the establishment of 3D models using CAD software. Those CAD models are then converted to STL format to be read by 3D printer. Afterwards, 3D printer configuration must be set up following designer requirements. Once 3D printing setup is completed, the building process can be commenced. Removal and clean-up are conducted after entire 3D printed parts are built. When the end-users require final 3D printed parts with smooth surfaces and (or) high mechanical strength, post-processing processes such as mechanical, chemical and electrochemical, thermal as well as hybrid methods can be further implemented by different means.

References

1. Nugroho, W., Dong, Y., Pramanik, A.: Chapter 4—3D printing composite materials: A comprehensive review. In: Low, I.-M., Dong, Y. (eds) Composite Materials: Manufacturing, Properties and Applications, pp. 65–115. Elsevier B.V. (2021)
2. Grimm, T.: The rapid prototyping process. In: Grimm T. (ed) User's Guide to Rapid Prototyping, Michigan, Society of Manufacturing Engineers (2004)
3. Youne, M.A.M.B.: Features of architecture computer aided design based on modeling of component types. Adv. Comput. **4**(1), 6–9 (2014)
4. Chang, K.H.: In: Product Design Modeling using CAD/CAE. Academic Press. Chapter 2, pp. 39–123. (2014)
5. Autodesk.: Autodesk Fusion with Netfabb: Additive manufacturing, design, and simulation. https://www.autodesk.com/products/netfabb/overview?term=1-YEAR&tab=subscription, Accessed 12 Jan 2025
6. Dassult Systemès.: Catia. https://www.3ds.com/products/catia, Accessed 12 Jan 2025
7. Software Advice.: Solid Edge Reviews. https://www.softwareadvice.com/simulation/solid-edge-profile/reviews/, Accessed 12 Jan 2025
8. Software Advice. Onshape Reviews. https://www.softwareadvice.com/construction/onshape-profile/reviews/, Accessed 12 Jan 2025
9. Software Advice.: Rhino 7 Reviews. https://www.softwareadvice.com/fashion-design/jewelry-modeling-profile/reviews/, Accessed 12 Jan 2025
10. Dassault Systèmes.: Software for your digital end2end manufacturing workflow. Dassault Systèmes. https://au.3dsystems.com/software#design, Accessed 12 June 2021
11. Markforged. Eiger.: The leading digital manufacturing software for driving business transformation. Markforged. https://markforged.com/software, Accessed 5 June 2021
12. Airwolf3D.Download APEX 3D-Printing Software. Airwolf3D printers. https://airwolf3d.com/3d-printing-software/, Accessed 18 June 2021
13. Khalil, S., Sun, W.: Biopolymer deposition for freeform fabrication of hydrogel tissue constructs. Mater. Sci. Eng. C **27**, 469–478 (2007)
14. Guo, C., Zhang, M., Bhandari, B.: Model building and slicing in food 3D printing processes: a review. Comprehens. Rev. Food Sci. Food Safety **18**, 1052–1069 (2019)
15. Derossi, A., Caporizzi, R., Ricci, I., Severini, C.: Critical variables in 3D food printing. In: Godoi, F., Bhandari, B., Prakash, S., Zhang, M. (eds.) Fundamentals of 3D Food Printing and Applications, pp. 41–91. Academic Press, Cambridge (2019)
16. Wang, J., Shaw, L.: Rheological and extrusion behavior of dental porcelain slurries for rapid prototyping applications. Mater. Sci. Eng. A **397**, 314–321 (2005)
17. Airwolf3D.: Online store for prosumer, educational and professional 3D printers, accessories and filament. Airwolf 3D printers. https://airwolf3d.com/shop/, Accessed 18 June 2021
18. Tan, H., Fang, Y., Zhong, C., Yuan, Z., Fan, W., Li, Z., Chen, J., Lin, X.: Investigation of heating behavior of laser beam on powder stream in directed energy deposition. Surface Coat. Technol. **397**, 126061 (2020)
19. Bai, Y., Chaudhari, A., Wang, H.: Investigation on the microstructure and machinability of ASTM A131 steel manufactured by directed energy deposition. J. Mater. Process. Technol. **276**, 116410 (2020)
20. Bai, Y., Shi, Z., Lee, Y. J., Wang, H.: Optical surface generation on additively manufactured AlSiMg0.75 alloys with ultrasonic vibration-assisted machining. J. Mater. Process. Technol. **280**, 116597 (2020)
21. Ni, C., Zhu, L., Zheng, Z., Zhang, J., Yang, Y., Yang, J., Bai, Y., Weng, C., Lu, W.F., Wang, H.: Effect of material anisotropy on ultra-precision machining of Ti-6Al-4V alloy fabricated by selective laser melting. J. Alloys Compounds **848**, 156457 (2020)
22. Zhang, J., Wang, H.: Micro-blasting of 316L tubular lattice manufactured by laser powder bed fusion. In: Proceedings of the 19th International Conference of the European Society for Precision Engineering and Nanotechnology EUSPEN (2019)

23. Zhang, J., Chaudhari, A., Wang, H.: Surface quality and material removal in magnetic abrasive finishing of selective laser melted 316L stainless steel. J. Manufact. Process. **45**, 710–719 (2019)
24. Wang, J., Zhu, J., Liew, P.J.: Material removal in ultrasonic abrasive polishing of additive manufactured components. Appl. Sci. **9**(24), 5359 (2019)
25. Guo, J., Song, C., Fu, Y., Au, K.H., Kum, C.W., Goh, M.H., Ren, T., Huang, R., Sun, C.N.: Internal surface quality enhancement of selective laser melted inconel 718 by abrasive flow machining. J. Manufact. Sci. Eng. **142**(10), 101003 (2020)
26. Galantucci, L.M., Lavecchia, F., Percoco, G.: Experimental study aiming to enhance the surface finish of fused deposition modeled parts. CIRP Ann. **58**(1), 189–192 (2009)
27. Havenga, S., De Beer, D.J., Van Tonder, P.J.M., Campbell, R.I.: Effectiveness of acetone post-production finishing on entry level FDM printed ABS artefacts. In: RAPDASA Conference Proceedings, pp. 4–6. (2015)
28. Himmer, T., Nakagawa, T., Anzai, M.: Lamination of metal sheets. Comput. Indus. **39**(1), 27–33 (1999)
29. Iqbal, F., Alam, Z., Khan, D.A.: In: Iqbal, F., Alam, Z., Khan, D.A. (eds) Additive Manufacturing and Post-Processing: An Introduction in Post-Processing Techniques for Additive Manufacturing, CRC Press, Florida, US, pp. 1–20. (2024)
30. Tillmann, W., Schaak, C., Nellesen, J., Schaper, M., Aydinöz, M.U., Hoyer, K.P.: Hot isostatic pressing of IN718 components manufactured by selective laser melting. Addit. Manufact. **13**, 93–102 (2017)
31. Hur, J., Lee, K., Kim, J.: Hybrid rapid prototyping system using machining and deposition. Comput.-Aided Design **34**(10), 741–754 (2002)

Chapter 3
3D Printing of PU Composites

Abstract This chapter highlights the fabrication of polyurethane (PU) composite structures using 3D printing that involves purification, dehumidification and mixing processes in the preparation step. Purification and dehumidification aim to remove the impurities and minimise the moisture content of PU and PU composites. Meanwhile, the mixing process, consisting of solution mixing, melt mixing and the combination of both methods, is implemented to obtain a homogeneous mixture. The mixture acts as feedstock materials that are processed by 3D printing to produce PU composite samples. Those samples are then tested to determine their mechanical, thermal and electrical properties.

Keywords 3D printing · PU composites · Purification · Dehumidification · Mixing process · Microstructures

3.1 Preparation of PU Composites

In general, 3D printed PU composites are generally prepared using three fundamental steps, namely purification, dehumidification and mixing. Purification aims to separate unwanted elements from the mixture [1]. Dehumidification assists in decreasing the moisture content of a material. Meanwhile, mixing represents the merger of two or more materials into one mass. This section elaborates the preparation of PU composites reinforced with typical nanofillers such as halloysite nanotubes (HNTs), carbon nanotubes (CNTs) and silver nanoparticles (AgNPs) in order to create a comprehensive study on 3D printed PU composites.

Evaporation, washing and drying are typically applied to purify and lower the moisture content of PU composites [2]. This research employed an evaporator and vacuum filter to get rid of the pulp from organosolv lignin and split organosolv solution from the extracted lignin in the lignin extraction process. During the evaporation process, the solvent in the organosolv solution was eliminated. On the other hand, the washing process was performed to remove the excess amount of formic acid from the lignin. Afterwards, the lignin was recovered via an oven-drying process.

W. T. Nugroho et al., *Thermoplastic Polyurethane (TPU)/Halloysite Nanotube (HNT) Nanocomposites*, Engineering Materials, https://doi.org/10.1007/978-981-95-0541-8_3

Some of PU composites require drying and sieving to limit moisture content and impurities [3]. This study depended on the selective laser sintering (SLS) technique with the material selection of PU, CNTs and polyamide 12 in powder form as feedstock materials to create composite parts. A Bucher funnel was employed to screen the suspension of PU/CNT mixture with reduced pressure at material preparation step. Subsequently, dehumidification (drying) took place in a vacuum oven at 50 °C for 24 h in order to minimise the moisture content of PU/CNT mixture. The mixture was then sieved to eliminate particles with undesirable sizes [3]. This study applied two purification techniques of filtration and sieving along with the dehumidification method for drying. Gan et al. [4] also utilised drying and sieving techniques to prepare PU/single-walled carbon nanotube (SWCNT) nanocomposites in powder form via SLS. When the diameter/size of nanofillers such as HNTs is relatively uniform, such a sieving technique is not applicable anymore [5]. As a result, it may reduce the preparation steps of PU/HNT nanocomposites when compared with PU/CNT nanocomposites. PU pellets and HNT particles were then dehumidified in a dryer at 80 °C for 4 h prior to the mixing process.

The mixing method affects filler dispersion in the matrices to improve composite properties, consisting of mechanical strength, thermal and electrical conductivities. In relation to the fillers, different polymers, fibres and nanoparticles can be utilised as the reinforcements for PU matrices. On the other hand, the nature of raw materials should be taken into account in used mixing technique in order to mix constituent materials for the preparation of composite materials [6]. This section covers three typical mixing methods, namely solution mixing, melt mixing and their combination.

Solution mixing is usually employed to prepare PU composites prior to 3D printing via direct ink writing (DIW), SLS and stereolithography (SLA) [2, 4, 7]. Such a technique was applied to mix PU with other materials for the manufacture of PU composites using DIW [7]. In this study, PU was combined with urea–formaldehyde (UF), formaldehyde, acetone and water to create shell materials, in which polyolefin-based structural fibres were chosen as the core. Subsequently, shell materials were coated onto core material surfaces using DIW in order to manufacture composite structures. This coating aimed to shield core materials and enhance chemical bonding between fibres and cement matrices in possession of the hydrophilicity to fibre surfaces. Gan et al. [4] also used solution mixing to combine PU and SWCNTs in order to produce composite powders applicable to SLS. Firstly, SWCNTs were pre-dispersed via wet ball milling. They were then mixed with polyvinylpyrrolidone (PVP). This mixture was rotated at 300 rpm to avoid the shear stress in order to break up SWCNT particles. The dispersion obtained was placed into a beaker with the addition of anhydrous ethanol, which was followed by ultrasonication to prepare stable SWCNT dispersion. PU powders were then gradually added by means of continued mechanical stirring for 2 h [4].

The preparation of feedstock materials used for SLA, derived from renewable materials (e.g. natural lignin), graphene nanoplatelets (GNPs) and photocurable PU, typically involves solution mixing. Lignin is received from oil palm empty fruit bunch (EFB) fibres using an organosolv process. A homogeniser was utilised to blend between the lignin and GNPs for 30 min. Photocurable PU was then added to this solution. The whole process took approximately 10 min until the particles were homogeneously dispersed [2].

The melt mixing technique needs a mixing device, which is referred to as an extruder or compounder to combine raw materials where such a mixing process typically occurs above glass transition temperature or melting temperature of polymers. Thermoplastics and elastomers are commonly used as the matrices, while nanoparticles and copped materials are selected as the fillers [8, 9]. An extruder used to create composite filaments via fused deposition modelling (FDM) is known as a filament extruder. This kind of extruder can be divided into two different types, namely single-screw and twin-screw extruders. Their advantages and drawbacks are listed in Table 3.1.

According to extruder specifications presented in Table 3.1, a single-screw extruder can be better option for filament fabrication with small outputs, low-cost investment and medium filament quality. Conversely, a twin-screw extruder is applicable when the users require good particle/fibre dispersion through more effective shearing mechanism in composite filaments.

Melt mixing with a twin-screw extruder on a small scale was previously employed to fabricate the filaments from PU and MWCNTs [10]. This study adjusted extruding parameters, consisting of melting temperature in range of 219-228 °C, screw speed of 300 rpm and screw length-to-diameter (L/D) ratio of 36. Other studies [11, 12] also utilised a twin-screw extruder to mix raw materials based on PU and polylactic acid (PLA). The main distinction was that the former selected an ordinary PU as the matrices and implemented a twin-screw extruder to produce granulated composites. Such composite granules were then melt-compounded by a single-screw extruder to manufacture FDM filaments. The latter mixed PU with PLA to create composite parts with an evident shape memory behaviour.

A single-screw extruder could be used to perform a melt mixing process in order to produce the filaments from PU and cork [13]. PU pellets were milled and sieved

Table 3.1 Advantages and drawbacks of single-screw and twin-screw filament extruders [6]

Parameter	Single-screw extruder	Twin-screw extruder
Capital cost	Low	High
Heat of friction	High	Low
Usage procedure	Easy	More complicated
Compounding ability	Low	High
Assemble and disassemble	Easy	Difficult
Energy consumption	Low	High
Stay time	Long	Short
Rotary speed	Low	High
Conveying efficiency	Low	High
Versatility	Low	High
Process productivity	Low	High

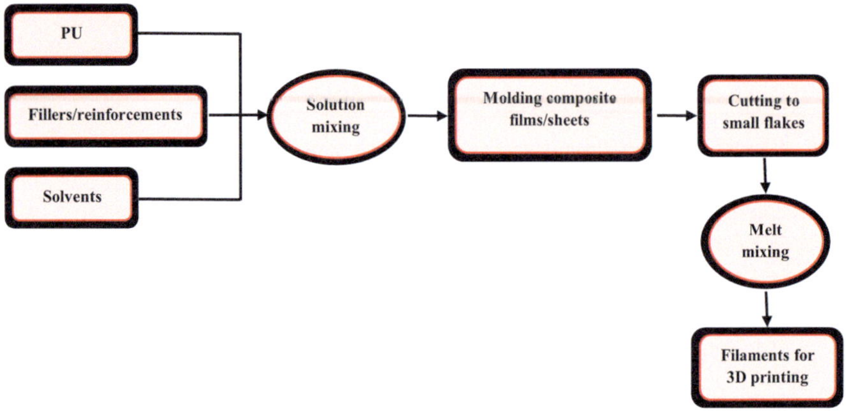

Fig. 3.1 Combined solution mixing and melt mixing process [6]

to prepare PU powders before those materials were blended with cork powders and further dried in an oven in order to minimise the moisture content of the mixture. Afterwards, the mixture was processed via melt mixing and extrusion to yield the filaments used for subsequent 3D printing via FDM.

In a special case, raw materials are not able to be directly combined using either solution mixing or melt mixing. As such, solution mixing is employed to create composite films that are cut into small flakes. Those flakes are then combined utilising melt mixing method in an extruder to manufacture composite filaments, as shown in Fig. 3.1.

In general, the combination method of solution mixing and melt mixing is applied to fabricate feedstock materials for 3D printed PU and PU composites manufactured via FDM [8]. This study implemented such a method to combine PU and zeolitic imidazolate framework (ZIF-8) to produce composite filaments. First of all, PU filaments were dissolved in dimethylformamide (DMF) and then stirred to make a PU solution. ZIF-8 powders were then added in acetone by bath sonication to create a metal–organic framework (MOF) solution. Afterwards, MOF and PU solutions were mixed, and acetone was eliminated via rotary evaporation. This mixture was poured onto a dish and placed on a hot plate for solvent removal. The following process was to dry the composites in a vacuum oven in order to create composite films. Such films were then chopped into small flakes and loaded into a filament extruder, subjected to a melt mixing process by extrusion to manufacture composite filaments.

Xiang et al. [14] implemented the combination of solution mixing and melt mixing methods to prepare a mixture derived from PU, silver nanoparticles (AgNPs) and CNTs. In this study, PU was in powder form instead of filaments. Nanocomposite sheets resulting from solution mixing were chopped into small granules, underwent a melt mixing in a single-screw extruder and were finally extruded into composite filaments via FDM.

The melt mixing is less complicated, inexpensive and less time-consuming compared to solution mixing. However, melt mixing has a low level of homogeneity for filler dispersion, in which filler dispersion can significantly affect composite properties like mechanical strength, thermal and electrical conductivities. Therefore, a combination of solution mixing and melt mixing methods is required to tackle the drawback of melt mixing process alone. Nonetheless, such combined methods rely on the use of solvents and additional apparatuses to complete the mixing step.

3.2 Microstructures and Properties of PU Composites

PU with unique characteristics between rubbers and plastics has been widely used as feedstock materials for 3D printing in order to manufacture useful components such as flexible sensors and actuators [15, 16]. Even though 3D printed PU can meet user's demands relating to complex geometry, its mechanical strength and functionality still become a major obstacle for widespread applications. Hence, other materials like particles, fibres and polymers as additives are mixed with PU to overcome its demerits. This combination results in unique microstructures that influence material properties of final PU composite parts.

The presence of nanoparticle reinforcements aims to improve properties of PU nanocomposites since they are very cheap and become facile to be combined with PU through the mixing processes. Such nanoparticles can enhance tensile strength, elastic modulus, electrical and thermal conductivities [4, 5, 9, 10, 17]. The addition of 1 wt% HNTs increased both tensile and flexural strengths of PU/HNT nanocomposites by 184% and 293% respectively, as opposed to those of neat PU. On the other hand, the presence of 1 wt% MWCNTs enhanced those of PU/MWCNT nanocomposites up to 112% and 150% accordingly [5]. Therefore, the reinforcement of HNTs resulted in greater tensile strength and flexural strength on PU nanocomposite samples than those reinforced with MWCNTs.

Xiang et al. [9] also found that the incorporation of 3 wt% MWCNTs in PU matrices improved tensile strength and Young's modulus of modified (M-CNT/PU printed) and unmodified (CNT/PU) nanocomposites by 49 and 39% respectively, as shown in Fig. 3.2. This improvement could be related to positive effect, arising from the mild disentanglement and the alignment of CNTs due to shearing effect in a single-screw extruder, as well as the printer nozzle. The disentanglement reduced particle agglomeration, which promoted filler dispersion and elongation of 3D printed material samples. The alignment of layers in the FDM process yielded laminate structures, especially in the longitudinal orientation under tensile loading. This alignment was beneficial to enhance elastic modulus and tensile strength.

The filler-matrix interaction is a critical factor in relation to filler dispersion. Great efforts have been made to improve such a critical interaction with non-covalent surface modification of MWCNTs by involving 1-pyrenecarboxylic acid (PCA) [18]. Excellent interaction can be obtained if filler surfaces stick well to those of the matrices in order to make an excellent unit cell of composite materials without

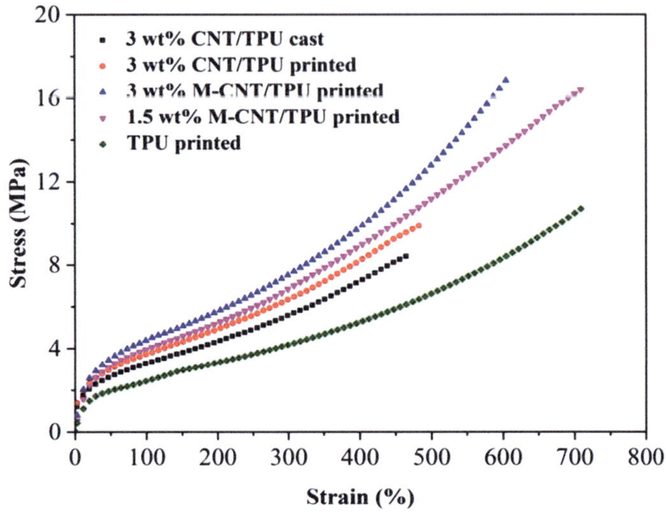

Fig. 3.2 Stress–strain curves of PU/MWCNT nanocomposites [9]

any presence of overlaps, voids or defects. Two methods can be implemented to improve interfacial bonding of composites, namely covalent and non-covalent modifications. Nevertheless, covalent modification has serious deficiency since it can often damage intrinsic structures of nanofillers. As a result, it deteriorates mechanical and electrical properties of nanofillers. Conversely, non-covalent modification enables to enhance nanofiller dispersion without the destruction of their intrinsic structures [9]. However, the presence of nanoparticles does not always enhance the properties of polymer nanocomposites. When MWCNT content reached 5 wt%, the tensile strength of nanocomposites declined by 72.24% [10]. It could arise from weak interfacial strength of interlayer bonds at such a filler content, which induced stress concentration sites or detrimental defects.

The impact of filler addition on the electrical conductivity of 3D printed nanocomposites was reported by Gan et al. [4]. This study utilised SWCNTs to wrap PU powders, which was known as SWCNT-wrapped-PU (SWCNT-w-PU). It aimed to increase the electrical conductivity of 3D printed nanocomposites in a solution mixing process [19]. SLS technique was applied to process such a solution in order to build porous surface macrostructures combined with conductive-segregation microstructures used for a flexible piezoresistive sensor [4]. Scanning electron microscopic (SEM) and transmission electron microscopic (TEM) images were observed to characterise nanocomposite powders shown in Fig. 3.3.

SWCNT-w-PU nanocomposite powders possessed irregular, but relatively uniform shapes that were identical to those of PU powders shown in Fig. 3.3a. The wrapping structures of SWCNT-w-PU nanocomposites played an important role in manufacturing conductive-segregation network structures in the matrices, as indicated by the absorption of individually dispersed SWCNT particles on PU surfaces

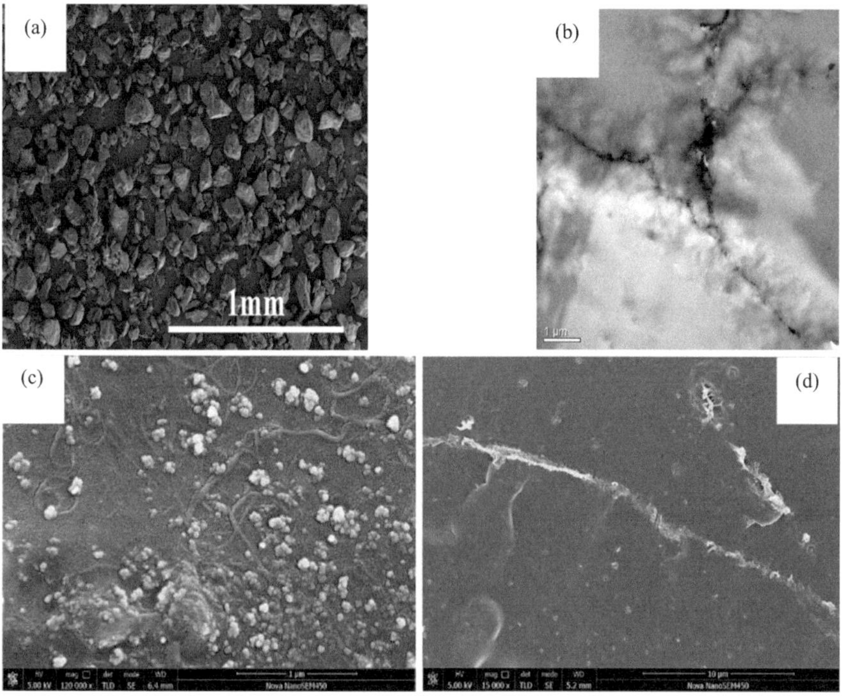

Fig. 3.3 **a** SEM images of 0.2 wt% SWCNT-w-PU nanocomposite powders, **b** TEM image of 3D printed PU/0.2 wt% SWCNT nanocomposites, **c** SEM image of 0.2 wt% SWCNT-w-PU nanocomposite powders surface and **d** SEM image of 3D printed PU/0.2 wt% SWCNT nanocomposites [4]

illustrated in Fig. 3.3c. Like other 3D printing techniques, SLS manufactures a part by additively depositing feedstock materials layer by layer. It signifies that throughout the depositing process, interfacial PU melt flows across interfacial SWCNTs, which then created coalescences among contiguous polymer particles. PU melt has a high viscosity, which induces near-zero shear stress of melt flow that cannot significantly alter the morphology of dispersed SWCNT particles in 3D printing process. As such, those particles remained at the particle boundaries and built conductive segregated networks. The formation of a conductive path of SWCNT particles in PU matrices is depicted in Fig. 3.3b and d.

The effect of SWCNT content on electrical conductivity of composite samples according to filler wrapped PU powders was compared with that emerging from other reinforcements like MWCNTs and graphene (GE). This wrapping phenomenon resulted in different conductivity impacts for such particles. It could be associated with the electrical percolation threshold of resulting composites. Three types of composites, consisting of PU/SWCNT composites, PU/MWCNT composites and PU/GE composites, possess conductive percolation thresholds of 0.05, 0.2 and 1 wt% respectively, as shown in Fig. 3.4. The incorporation of SWCNTs up to 0.05

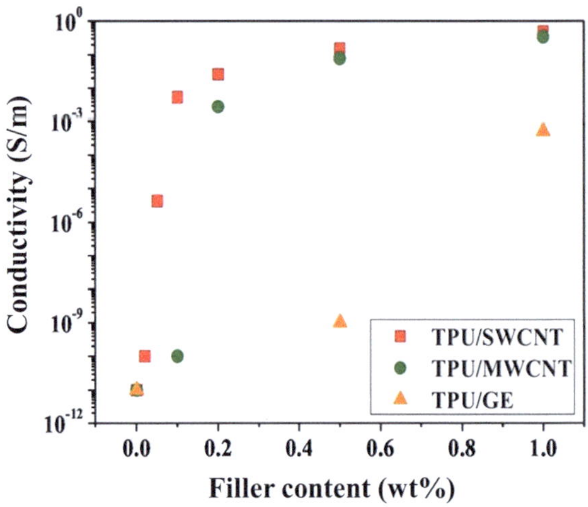

Fig. 3.4 Electrical conductivity of PU composites with different fillers using SLS [4]

wt% significantly increased the electrical conductivity of PU/SWCNT composite samples. At low filler contents below 1 wt%, the electrical conductivity of PU/SWCNT composites was slightly greater than that of PU/MWCNT composites despite being much higher than PU/GE composites.

Increasing MWCNT content in PU matrices can enhance the thermal conductivity of 3D printed composites [10]. Even though composite samples produced relatively low thermal conductivity values in range of 0.22–0.55 W/m·K, the presence of 5 wt% MWCNTs enhanced thermal conductivity of PU/MWCNT composite samples by 100% when compared with 3D printed neat PU [10].

Table 3.2 summarises nanoparticles employed in 3D printed PU composites and corresponding improved composite properties. When compared with AgNPs and SWCNTs, MWCNTs can be considered as effective reinforcements in PU matrices since such nanoparticles enable to enhance tensile strength, elastic modulus, electrical and thermal conductivities. On the other hand, FDM is employed to process PU composites more frequently than other 3D printing methods. It might be supported by the capability of FDM to handle various materials, which allows FDM to concurrently combine different materials like PU and PU composites.

Cellulose nanofibrils (CNFs) are also very popular as PU reinforcements to improve mechanical properties of composite parts such as tensile strength, elasticity modulus and toughness [20]. Mohan et al. [20] employed CNFs derived from empty fruit bunch (EFB) as the reinforcements in PU matrices to enhance the tensile strength of 3D printed composites. CNFs can be extracted from agricultural waste of palm oil mill with rich lignocellulose. Lignocellulose may be further processed to make nanocellulose as reinforcing fillers in polymer matrices [21]. Nonetheless, CNFs should be chemically treated to increase their compatibility with UV-curable and PU-based resins, which is considered as feedstock materials for digital light processing (DLP). Hence, the hydrophilicity of CNF surfaces must be adapted with polyethylene glycol (PEG) and reduced graphene oxide (rGO) in order to produce a

Table 3.2 Summary of enhanced properties in 3D printed PU composites reinforced by nanoparticles [6]

Method	Material system	Optimum filler content	Improvement in properties	References
FDM	• PU/HNT composites • PU/MWCNT composites	• 1 wt% of HNTs • 1 wt% MWCNTs	• Tensile strength and flexural strength • Tensile strength and flexural strength	[5]
FDM	• PU/MWCNT composites	• 3 wt% of MWCNTs	• Tensile strength, strain and gauge factor (GF)	[9]
		• 5 wt% of MWCNTs	• Tensile strength, Young's modulus and electrical conductivity	[17]
		• 5 wt% of MWCNTs	• Thermal conductivity	[10]
	• PU/MWCNT/AgNP composites	• 3 wt% of MWCNT/AgNP	• Elastic modulus	[14]
SLS	• PU/CNT composites	• 1 wt% of CNTs	• Electrical conductivity	[3]
SLS	• PU/SWCNT composites	• 0.05 wt% of SWCNTs	• Electrical conductivity	[4]

good mixture in combination with UV-curable PU resin. Figure 3.5 shows the micrographs of sample cross-section observed by a field emission scanning microscope (FE-SEM).

The successful cellulose defibrillation is presented in Fig. 3.5a where individual fibrils possessed a comparable range of 5–20 μm in average diameter [22]. Conversely, rGO turned to an unstable semi-transparent sheet material under TEM observation shown in Fig. 3.5b. Disordered and unwrinkled textures might come from non-removal of oxygen atoms and a high degree of exfoliation under oxidation. Figure 3.5c depicted the SEM image of PEG/CNF composites with respect to CNF dispersibility in PEG matrices. Meanwhile, Fig. 3.5d showed CNF visible traces under a thin layer of rGO [23]. The modified CNFs indicated interfacial agent stability while remaining as individual fibrils.

The dispersibility of modified CNFs in UV-cured PU was also investigated using tensile properties, as illustrated in Fig. 3.6. The addition of PEG and rGO with 1 wt% modified CNFs improved corresponding tensile strengths by 8.3% and 9.37% respectively [20]. It might be associated with the fact that PEG/CNF composites and CNF/rGO composites could facilitate PU crystallisation processes in their composite systems [24]. Graphene is deemed a well-known carbon-based nanofiller, in which its layer structures possess very large surface areas up to 2630 m^2/g [25]. Therefore, a lower loading of those nanofillers can improve composite properties owing to their high aspect ratios [26], as clearly seen in Fig. 3.6. PU-3%-CNF/PEG composites had higher tensile strength by 24% than neat PU. PU-3%-CNF/rGO composites

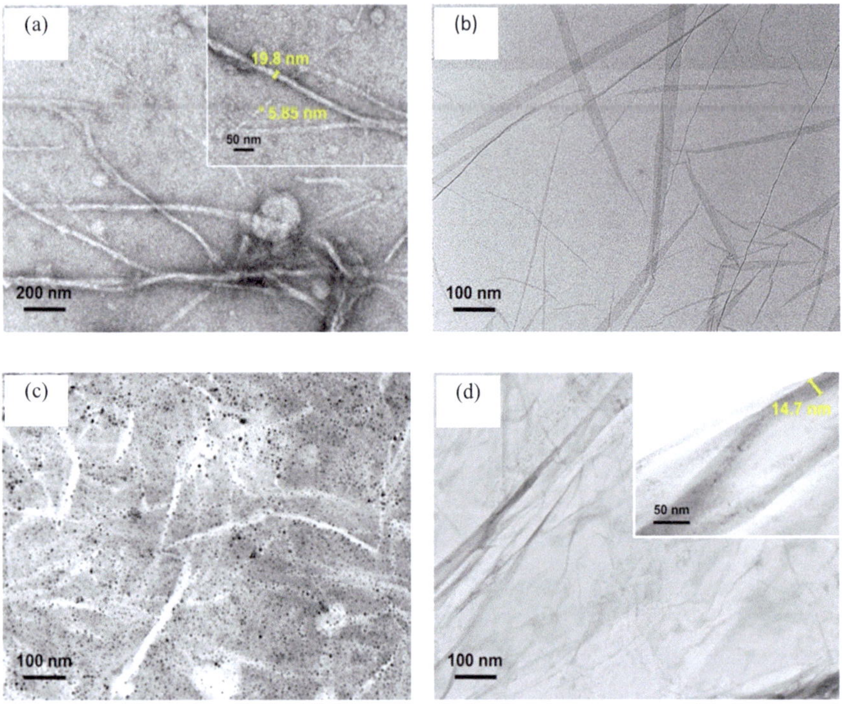

Fig. 3.5 Micrographs of **a** unmodified CNF, **b** rGO, **c** surface-grafted PEG/CNF composites and **d** surface-grafted CNF/rGO [20]

even enhanced tensile strength up to 11% and 37% when compared with those of PU-3%-CNF/PEG composites and PU accordingly.

The addition of PEG could ease the interaction between hydrophobic PU and hydrophilic CNFs. On the other hand, rGO provided better removal of hydrophilicity

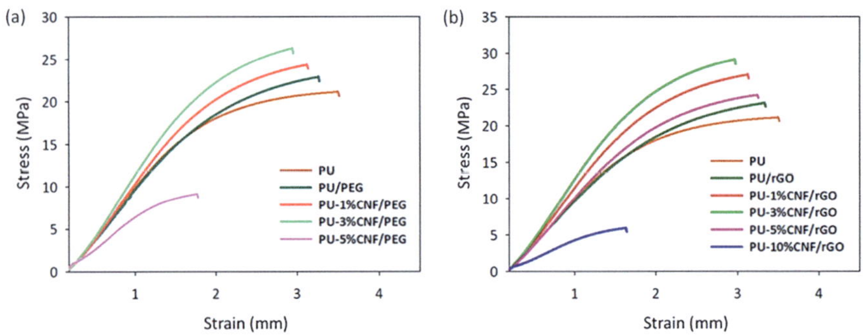

Fig. 3.6 Stress–strain curve of UV-cured PU and modified cellulose reinforced with **a** PEG and **b** rGO [20]

Fig. 3.7 FE-SEM images of 3D printed composite samples at different PANI contents: **a** 1 wt%, **b** 3 wt% and **c** 6 wt% [31]

because of a higher carbon density on its surfaces [27, 28]. rGO could overcome natural hydrophilicity of CNFs, enabling CNFs to be better dispersed within hydrophobic PU matrices [29]. Nonetheless, when modified CNF content reached 5 wt%, tensile properties declined due to resulting non-uniform stress transfer under tension loading. It signified that CNFs tended to aggregate in PU matrices and generated more interfacial voids to undermine the tensile strength of composite materials beyond the filler content of 5 wt% [30]. A higher CNF content could retard the curing process of photocurable PU resin and further affect 3D printed sample direction owing to insufficient curing.

Polyaniline (PANI) enables to increase the electrical conductivity of 3D printed composites via DLP, which may be utilised as sensors and antistatic materials in biomedical fields [31]. The impact of PANI content on 3D printed composites was investigated using FE-SEM images, as illustrated in Fig. 3.7. Increasing PANI content created a wider area of PANI embedded in PU matrices, which caused intermaterial aggregation. Such aggregation might be ascribed to strong hydrogen bonding, as well as London force between dipole–dipole and PANI chains [31–33]. PANI aggregation influenced the crosslinking between PU prepolymers [34]. Accordingly, the maximum PANI content in composites was fixed at 6 wt%.

On the other hand, PANI content above 6 wt% decreased the sensitivity of composite samples. The optimum conditions, namely the lowest resistance and the highest electrical conductivity of 3D printed composites, were reached at 6 wt% PANI, as indicated in Fig. 3.8. At this content level, the resistance of PU/PANI composite became approximately 8.57×10^4 times lower than that of pure PU. On the other hand, its electrical conductivity was 8.57×10^4 times higher than that of PU [31]. A summary of property enhancement in 3D printed PU composites reinforced with fibres and polymers is presented in Table 3.3.

3.3 Summary

Preparation, microstructures and properties of PU composites have been discussed in this chapter. Purification, dehumidification and mixing processes are required in the preparation step to fabricate feedstock materials derived from PU composites for 3D printing. FDM and SLS are typically used to process PU composites

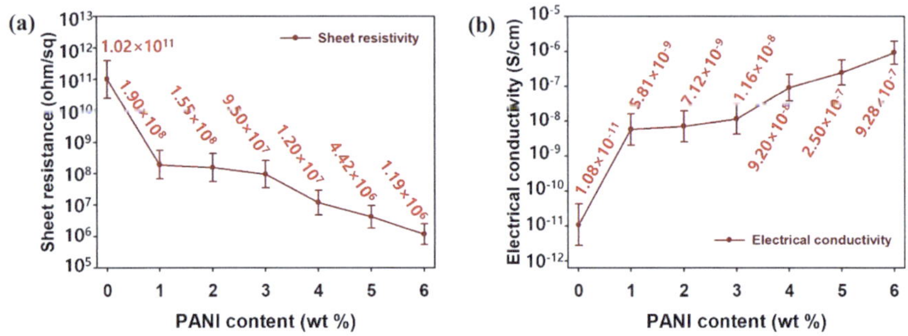

Fig. 3.8 (a) Resistance and (b) electrical conductivity of 3D printed composites [31]

reinforced with nanoparticles, such as HNTs, MWCNTs and SWCNTs, in order to manufacture 3D printed nanocomposite samples. The incorporation of 1 wt% of HNTs or MWCNTs can improve tensile strength and flexural strength. When the MWCNT content increased up to 5 wt%, it further enhanced tensile strength, Young's modulus, electrical and thermal conductivities. Meanwhile, the incorporation of 0.05 wt% SWCNT successfully increased the electrical conductivity of 3D printed PU/SWCNT composites. Overall, PU composites reinforced with fibres and polymers can be processed by various 3D printing techniques including FDM, DIW and DLP. The presence of 70% vol% PLA improved tensile strength and elasticity modulus, while the addition of 3 wt% CNFs may promote the compatibility of initial mixture solution with UV-curable PU resin and further improve tensile strength of 3D printed PU composites.

Table 3.3 Summary of enhanced properties in 3D printed PU composites reinforced with fibres and polymers [6]

Method	Material system	Optimum filler content	Improvement in properties	References
FDM	• PU/PLA	• 20 wt% of PLA	• Tensile strength	[11]
		• 70 vol% of PLA	• Tensile strength and elasticity modulus	[35]
		• 75 wt% of PLA	• Tensile strength	[36]
	• PU/PLA/graphene oxide (GO)	• 0.5 wt% of GO	• Tensile modulus and thermal stability	[37]
	• PU/hexagonal boron nitride (hBN)	• 40 wt% of hBN	• Thermal conductivity	[38]
DIW	• Polyolefin fibre/ urea–formaldehyde resin (UF) • Polyolefin fibre/PU resin	• 5 wt% of sodium silicate	• Interfacial toughness	[7]
DLP	• PU/CNF/reduced graphene oxide (rGO)	• 3 wt% of cellulose nanofibres (CNFs)	• Compatibility of composite solution with UV-curable PU resin and tensile strength at CNF contents below 3 wt%	[20]
	• PU/CNF/graphene nanoplatelet (GNP)	• 0.6 wt% of lignin/graphene	• Tensile strength and hardness	[2]
	• PU/Polyaniline (PANI) • PU/Graphene sheet (GS)	• 6 wt% of PANI • 2 wt% of GE	• Electrical conductivity	[31]
	• polylactic acid-polyurethane (PLA-PUA)/ trimethylolpropane trimethacrylate (TEGDMA)/graphene	• 37 wt% TEGDMA, 62 wt% PLA-PUA, 1 wt% Irgacure 819 and 0.5 wt% graphene • 3D printing with the layer thickness of 0.02 mm in *X*-axis direction	• Tensile strength and flexural strength	[39]

References

1. Armington, A.: Standard purification techniques. Annals of the New York Acad. Sci. **118**, 613–617 (2006)
2. Ibrahim, F., Mohan, D., Sajab, M., Bakarudin, S., Kaco, H.: Evaluation of the compatibility of organosolv lignin-graphene nanoplatelets with photo-curable polyurethane in stereolithography 3D printing. Polymers **11**, 1–12 (2019)

3. Li, Z., Wang, Z., Gan, X., Fu, D., Fei, G., Xia, H.: Selective laser sintering 3D printing: a way to construct 3D electrically conductive segregated network in polymer matrix. Macromolecular Mater. Eng. **1700211**, 1–10 (2017)
4. Gan, X., Wang, J., Wang, Z., Zheng, Z., Lavorgna, M., Ronca, A., Fei, G., Xia, H.: Simultaneous realization of conductive segregation network microstructure and minimal surface porous macrostructure by SLS 3D printing. Mater. Design **178**, 1–10 (2019)
5. Namathoti, S., Vakkalagadda, M.: Mechanical and shape recovery characterization of MWCNTs/HNTs-reinforced thermal-responsive shape-memory polymer nanocomposites. Polymers **15**, 710 (2023)
6. Nugroho, W., Dong, Y., Pramanik, A., Leng, J., Ramakrishna, S.: Smart polyurethane composites for 3D or 4D printing: general-purpose use, sustainability and shape memory effect. Compos. Part B Eng. **223**, 109104 (2021)
7. Eom, H., Won, J.: Selection of functional materials for surface-coating of structural synthetic fibres and development of a 3D-printing system for their manufacture. Compos. Struct. **249**, 1–7 (2020)
8. Evans, K., Kennedy, Z., Arey, B., Christ, J., Schaef, H., Nune, S., Erikson, R.: Chemically active, porous 3D-printed thermoplastic composites. ACS Appl. Mater. Interf. **10**, 15112–15121 (2018)
9. Xiang, D., Zhang, X., Li, Y., Jones, E., Zheng, Y., Wang, L., Zhao, C., Wang, P.: Enhanced performance of 3D printed highly elastic strain sensors of carbon nanotube/thermoplastic polyurethane nanocomposites via non-covalent interactions. Compos. Part B Eng. **176**, 107250 (2019)
10. Tzounis, L., Petousis, M., Grammatikos, S., Vidakis, N.: 3D printed thermoelectric polyurethane/multiwalled carbon nanotube nanocomposites: a novel approach towards the fabrication of flexible and stretchable organic thermoelectrics. Materials **13**, 1–16 (2020)
11. Leng, J., Wu, J., Zhang, J.: Preparation of thermoplastic polyurethane parts reinforced with in situ polylactic acid microfibers during fused deposition modeling: the influences of deposition-induced effects. ACS Indus. Eng. Chem. Res. **58**, 21476–21484 (2019)
12. Jing, X., Mi, H., Peng, X., Turng, L.: The morphology, properties, and shape memory behavior of polylactic acid/thermoplastic polyurethane blends. Polym. Eng. Sci. **55**, 70–80 (2014)
13. Gama, N., Ferreira, A., Timmons, A.: 3D printed cork/polyurethane composite foams. Mater. Des. **179**, 1–9 (2019)
14. Xiang, D., Zhang, X., Jones, E., Zhu, W., Zhou, Z., Shen, Y., Li, Y., Zhao, C., Wang, P.: Synergistic effects of hybrid conductive nanofillers on the performance of 3D printed highly elastic strain sensors. Compos. Part A: Appl. Sci. Manufact. **129**, 1–38 (2020)
15. Yuan, S., Li, S., Zhu, J., Tang, Y.: Additive manufacturing of polymeric composites from material processing to structural design. Compos. Part B **219**, 108903 (2021)
16. Wang, Y., Li, X.: 4D-printed bi-material composite laminate for manufacturing reversible shape-change structures. Compos. Part B **219**, 108918 (2021)
17. Kim, M., Jung, J., Jung, S., Moon, Y., Kim, D., Kim, J.: Piezoresistive behaviour of additively manufactured multi-walled carbon nanotube/thermoplastic polyurethane nanocomposites. Materials **12**, 1–24 (2019)
18. Legeza, O., Roder, J., Hess, B.: QC-DMRG study of the ionic-neutral curve crossing of LiF. Molecul. Phys. **101**, 2019–2028 (2009)
19. Lin, D., Nian, Q., Deng, B., Jin, S., Hu, Y., Wang, W., Cheng, G.: Three-dimensional printing of complex structures: Man made or toward nature? ACS Nano **8**, 9710–9715 (2014)
20. Mohan, D., Sajab, M., Kaco, H., Bakarudin, S., Noor, A.: 3D Printing of uv-curable polyurethane incorporated with surface-grafted nanocellulose. Nanomaterials **9**, 1–16 (2019)
21. Lamaming, J., Hashim, R., Sulaiman, O., Leh, C., Sugimoto, T., Nordin, N.: Cellulose nanocrystals isolated from oil palm trunk. Carbohydrate Polym. **127**, 202–208 (2015)
22. Chan, C., Chia, C., Zakaria, S., Sajab, M., Chin, S.: Cellulose nanofibrils: a rapid adsorbent for the removal of methylene blue. RSC Adv. **5**, 18204–18212 (2015)
23. Hou, M., Xu, M., Li, B.: Enhanced electrical conductivity of cellulose nanofiber/graphene composite paper with a sandwich structure. ACS Sustain. Chem. Eng. **6**, 2983–2990 (2018)

24. Rahmani, H., Najafi, S., Matin, S., Ashori, A.: Mechanical properties of carbon fiber/epoxy composites: effects of number of plies, fiber contents, and angle-ply layers. Polym. Eng. Sci. **54**, 2676–2682 (2013)
25. Parizi, T., Ebrahimi, G., Ezatpour, H.: Effect of graphene nanoplatelets content on the microstructural and mechanical properties of AZ80 magnesium alloy. Mater. Sci. Eng. A **742**, 373–389 (2019)
26. Park, Y., Qian, Y., Chan, C., Suh, T., Nejhad, M., Macosko, C., Stein, A.: Epoxy toughening with low graphene loading. Adv. Funct. Mater. **25**, 575–585 (2015)
27. Qu, P., Gao, Y., Wu, G., Zhang, L.: Nanocomposites of poly(lactic acid) reinforced with cellulose nanofibrils. BioResources **5**, 1811–1823 (2010)
28. Yang, Z., Li, X., Si, J., Cui, Z., Peng, K.: Morphological, mechanical and thermal properties of poly (lactic acid) (PLA)/cellulose nanofibrils (CNF) composites nanofiber for tissue engineering. J. Wuhan Univer. Technol. Mater. Sci. Edition **34**, 207–215 (2019)
29. Zainuddin, M., Raikhan, N., Othman, N., Abdullah, W.: Synthesis of reduced graphene oxide (rGO) using different treatments of graphene oxide (GO). In: Paper presented at IOP Conf. Series: Materials Science and Engineering, vol. 358, pp. 012046. (2018)
30. Venkatesh, A., Thunberg, J., Moberg, T., Klingberg, M., Hammar, L., Peterson, A., Muller, C., Boldizar, A.: Cellulose nanofibril-reinforced composites using aqueous dispersed ethylene-acrylic acid copolymer. Cellulose **25**, 4577–4589 (2018)
31. Joo, H., Cho, S.: Comparative studies on polyurethane composites filled with polyaniline and graphene for DLP-Type 3D printing. Polymers **12**, 1–11 (2020)
32. Jang, J.: Conducting polymer nanomaterials and their applications. Adv. Polym. Sci. **199**, 189–259 (2006)
33. Park, W., Kim, T., Huh, J., Kang, M., Lee, J., Yoon, H.: Anisotropic growth control of polyaniline nanostructures and their morphology-dependent electrochemical characteristics. ACS Nano **6**, 7624–7633 (2012)
34. Demir, M., Menceloglu, Y., Erman, B.: Aggregation of fillers blended into random elastomeric networks: theory and comparison with experiments. Macromolecular Chem. Phys. **207**, 1515–1524 (2006)
35. Shi, X., Chen, B., Tuo, X., Gong, Y., Guo, J.: Study on performance characteristics of fused deposition modeling 3D-printed composites by blending and lamination. J. Appl. Polym. Sci. **e32495**, 1–7 (2020)
36. Tao, Y., Shao, J., Li, P., Shi, S.: Application of a thermoplastic polyurethane/polylactic acid composite filament for 3D-printed personalized orthosis. Mater. Tehnol. **53**, 71–76 (2019)
37. Chen, Q., Mangadlao, J., Wallat, J., Leon, A., Pokorski, J., Advincula, R.: 3D printing biocompatible polyurethane/poly(lactic acid)/graphene oxide nanocomposites: anisotropic properties. ACS Appl. Mater. Interf. **9**, 4015–4023 (2017)
38. Liu, J., Li, W., Guo, Y., Zhang, H., Zhang, Z.: Improved thermal conductivity of thermoplastic polyurethane via aligned boron nitride platelets assisted by 3D printing. Composite Part A **120**, 140–146 (2019)
39. Feng, Z., Li, Y., Hao, L., Yang, Y., Tang, T., Tang, D., Xiong, W.: Graphene-reinforced biodegradable resin composites for stereolithographic 3D printing of bone structure scaffolds. Hindawi J. Nanomater. **2019**, 1–14 (2019)

Chapter 4
3D Printing of Re-Entrant Structures

Abstract This chapter focuses on the manufacturing of re-entrant (RE) structures made from polyurethane (PU) composites using 3D printing. Four types of structural models are generally selected, consisting of conventional honeycomb, RE honeycomb, tetrachiral and arrowhead structures. It is then followed by manufacturing techniques such as material extrusion (ME), material jetting (MJ), powder bed fusion (PBF) and vat photopolymerisation (VP) to fabricate such structures. Modelling approaches are also involved in order to create the best models in order to produce 3D printed structures with the desired characteristics. Finally, the properties of those RE structures, comprising Poisson's ratio (PR), compressive properties, energy absorption, fatigue, shape memory behaviour, electrical conductivity and piezoresistance, are discussed to provide valuable information in relation to 3D printed RE structures.

Keywords 3D printing · RE structures · Modelling approach · Multifunctional properties

4.1 Special Structural Design

Re-entrant (RE) structures with a unique construction made up from a number of well-designed cells to possess lightweight and high-strength properties due to their high stiffness/strength to mass ratio. They are also called auxetics because of resulting negative Poisson's ratio (NPR), reflected by its ability to demonstrate transverse expansion when subjected to uniaxial tension and transverse contraction under uniaxial compression [1]. RE structures were first proposed with remarkable shear stiffness by Lakes [2]. This finding has motivated to create and develop a variety of auxetic structures over the decades. In general, such efforts are mainly focused on structural design and manufacturing, material characterisation and modelling framework of RE structures. Finite element analysis (FEA) is one of the most popular approaches, which is applicable to create RE models in order to estimate their auxetic behaviour characterised by NPR and other critical multifunctional properties such as energy absorption, strain recovery and piezoresistance [3–5]. Initially, successful

W. T. Nugroho et al., *Thermoplastic Polyurethane (TPU)/Halloysite Nanotube (HNT) Nanocomposites*, Engineering Materials, https://doi.org/10.1007/978-981-95-0541-8_4

development of such models in 3D objects deals with the obstacle owing to their complex geometries and a lack of advanced manufacturing techniques and methods. This problem has been ultimately tackled with the presence of additive manufacturing (AM) technologies.

RE structure is a typical structure constructed by curves controlled by discrete points in space. Compared with other types of RE structures, conventional honeycomb, RE honeycomb, chiral and arrowhead structures are commonly used by researchers since chiral and arrowhead structures exhibit excellent auxetic features. On the other hand, conventional honeycomb and RE honeycomb are usually used as reference models. Figure 4.1 displays the unit cells of such structures where *H*, *L*, θ and *t* represent the height of vertical wall, the length of side cell, internal angle and cell thickness of all structures. Besides, *r*, L_1 and L_2 are cylinder radius of chiral structures and the lengths of two respective inclined cell ribs in arrowhead structures respectively.

Conventional honeycombs are made from hollow cells constructed between thin vertical cells, which are commonly seen in columnar and hexagonal shape illustrated

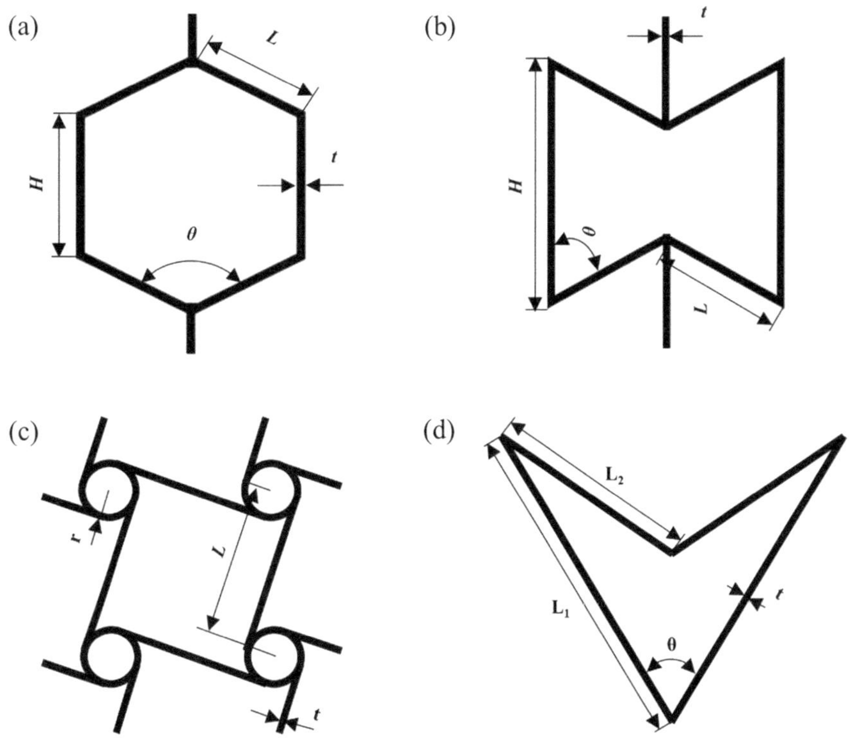

Fig. 4.1 Typical unit cells of **a** honeycomb, **b** RE honeycomb, **c** tetrachiral and **d** arrowhead structures [9]

in Fig. 4.1a. As such, honeycombs have a low density, as well as relatively high out-of-plane compressive and shear properties [6]. As inspired by a bee honeycomb, such structures are applicable in widespread applications such as mechanical engineering, architecture, chemical engineering, nanofabrication, biomedicine and transportation [7]. Furthermore, they can be considered as a basic model used for a regular lattice structure with a positive Poisson's ratio since the regular honeycomb expands with compressive loads. This behaviour is in contrast with RE auxetic structures that expand in all directions when subjected to tensile loads [8]. Accordingly, superior honeycombs are a crucial reference model for the ultimate upscaling and development of RE structures.

A sandwich structure derived from honeycomb cells was manufactured to investigate its ability as an impact absorber for automotive applications [10]. Critchley et al. [11] utilised two types of conventional honeycomb structures with internal angles of 105° and 120° in order to evaluate their ability to reduce the explosion and transmitted pressure. In contrast, conventional honeycomb patterns were incorporated as the sample core to understand their mechanical integrity and recovery capability prior to ultimate strain [12].

Generally speaking, "re-entrant" term is used to illustrate a directed inward behaviour or a negative angle, as depicted in Fig. 4.1b. The design of unique geometry of RE honeycombs enables these structures to generate NPR, which is reflected by their ability with transverse expansion under uniaxial tension according to Fig. 4.2a. Structural deformation is mainly induced by the re-alignment of cell ribs even though the deflection and axial deformation of cell ribs must also be considered among different mechanisms with respect to an auxetic behaviour [14]. RE honeycomb cells are usually highly anisotropic when compared with conventional counterparts [14]. As a result, RE honeycomb cells commonly possess higher transverse Young's modulus and shear modulus, as opposed to those of conventional honeycombs [15].

RE structures are typically made from lattice cell arrays so that an improved design becomes achievable based on their behaviour by optimising individual lattice cells. In recent years, the formation of novel shapes can be created from original cells in order to enhance mechanical properties with excellent material performance. In relation to this, a hybrid design may be feasible by combining some existing topographical behaviours to create a new configuration with some special effects not being obtained by a classical configuration [16, 17]. Additionally, the use of reinforcing ribs in hybrid designs is allowed to increase their stiffness [18].

3D models of RE structures are usually required since they cannot be easily evaluated by only using a regular array of 2D in-plane design. The rotational change, hybrid design, as well as part combination with different size methods may be applied to create 3D structures. Further development from 2D to 3D structures, along with mechanical properties of these structures, can be holistically evaluated according to more practical and accurate designed structures [19].

A chiral structure comprises a central cylinder and several tangentially attached ligaments, as illustrated in Figs. 4.1c and 4.2b. The basic unit of chiral structure can be divided into two classifications, namely "left-handed" or "right-handed" formation that possesses a chiral or anti-chiral characteristic. The anti-chiral nodes are attached

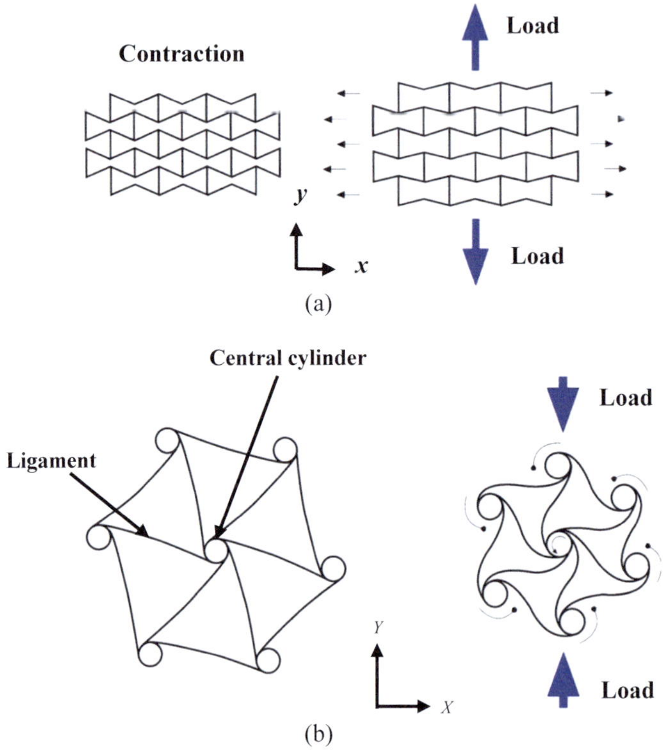

Fig. 4.2 **a** Reaction of RE structure under tensile loading in the *y*-direction and **b** schematic diagram of deformation mechanism for hexachiral structure in the *y*-direction [13]

on the same side of connecting ligaments leading to symmetrical structures. The cylinders enable to rotate under the applied load to chiral structures with flexible ligaments to achieve ligament folding or unfolding. It depends on geometric features of the structures that may make NPR as close as possible to -1 [20]. When the periodic chiral is made, the constraints of rotational symmetry must stand. The number of ligaments attached to each node should be equal to the order n of rotational symmetry [21]. Consequently, only five structural models exist including trichiral, anti-trichiral, tetrachiral, anti-tetrachiral and hexachiral structures [20, 21]. Moreover, a meta-chiral structure can be designed under relaxed constraints [21].

A meta-chiral structure possesses both chiral and anti-chiral characteristics. Some nodes of this structure are connected to the same side of the ligaments when compared with the others on the opposite sides. A study on a tetrameta-chiral model with rectangle nodes was conducted by Grima [21]. It was reported that this system was highly anisotropic with NPR beyond the isotropic limit of -1. Furthermore, Poisson's ratio depends on the angle between the ligament and nodes, as well as different aspect ratios.

With the difference from RE structures, Poisson's ratio of chiral structures depends on their structural angle. Chiral structures have the ability of winding themself when subjected to compression, leading to their Poisson's ratio beyond a significant range of strain approximately 25% [22]. The stiffness increase in cylinder-ligament systems comes from increasing the number of ligaments per node [20]. In general, chiral honeycombs possess much larger moduli when compared with anti-chiral counterparts due to the same number of available ligaments [20] notwithstanding opposite results obtained elsewhere [23]. The presence of cylinders usually reduces the in-plane stiffness of these structures in contrast with RE honeycombs.

Figure 4.1d demonstrates an arrowhead structure, which was first introduced by means of topological optimisation of RE structures [24]. Four critical parameters in arrowhead structures include cell wall thickness, the length of two different inclined cell ribs, as well as the internal angle between two cell ribs. A structural model was established to assess design parameters on compressive strength when subjected to quasi-static, high-velocity and low-velocity loads [25]. These results indicated that specific energy absorption (SEA) was in a proportional relationship with the square of impact velocity. A 3D arrowhead linkage auxetic structure was designed by intersecting 2D structures [26]. Furthermore, with the variation of structural geometries, it may be developed in order to calculate NPRs in all directions.

A jounce bumper has been modelled using arrowhead structures via both numerical and experimental approaches [27]. Polyamide and PU were considered as a stuffer and a tendon material to manufacture a jounce bumper using vat photopolymerisation (VP). This study revealed that join bumper could effectively perform its function under various loading conditions [27]. Damping properties of arrowhead structures were investigated by Ma et al. [28], which reported that the desired stiffness and damping capacity could be achieved by optimal inclined angle and stacking layer thickness. An empirical formula was developed by Zhao et al. [29] to evaluate different deformation modes in relation to critical impact velocity. It was noted that dynamic crushing strength was only influenced by relative density and impact velocity. The influence of geometrical parameters on elastic modulus and NPR of 3D arrowhead structures was assessed using FEA [30]. Numerical results showed that mechanical performance of 3D arrowhead structures was significantly affected by arrowhead angles.

4.2 Manufacturing Techniques

Unique deformation characteristics of RE structures with NPR have motivated researchers to investigate this field intensively. However, geometric complexity of those structures restricts further progress with research attempts limited to theoretical approaches only. The primary issue to resolve lies in how to fabricate RE hollow structures that are difficult to be manufactured with conventional manufacturing such as extrusion and injection moulding. Owing to technological advances, more sophisticated additive manufacturing (AM), called 3D printing, becomes widely available,

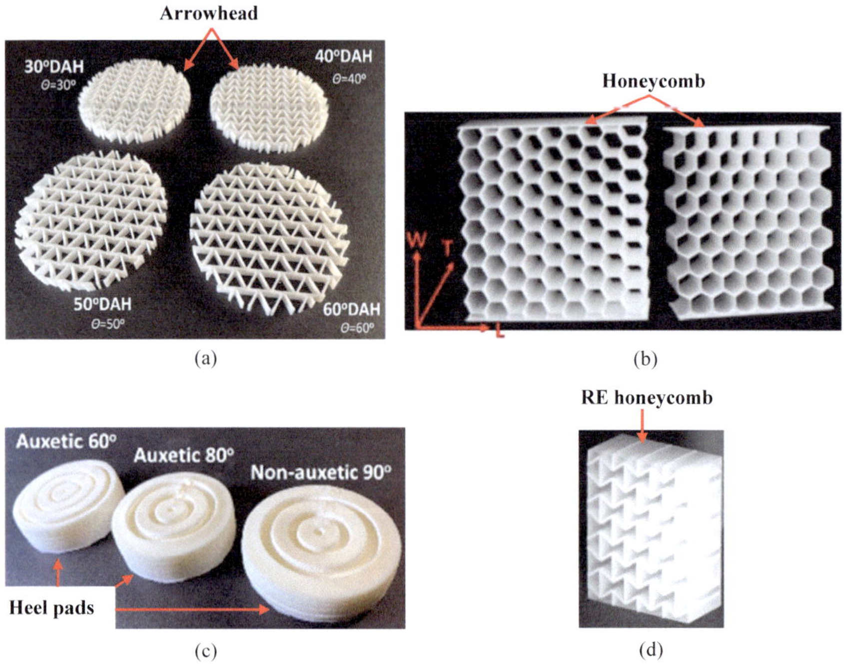

Fig. 4.3 3D printed structures manufactured by **a** FDM [31], **b** PBF [34], **c** VP [37] and **d** MJ [32]

which can be classified as a very new rapid prototyping technology, allowing to create complex structures from raw materials in a layer-by-layer manner [27]. In recent years, AM development has become relatively mature and commercially available. Among various AM techniques, material extrusion (ME) [12, 19, 31], material jetting (MJ) [32, 33], powder bed fusion (PBF) [34–36] and VP [3, 37, 38] are widely used to build RE structures. AM-based structures like conventional honeycomb, re-entrant honeycomb, chiral and arrowhead structures are depicted in Fig. 4.3.

Material extrusion is one of the most popular AM techniques according to the extrusion of feedstock materials through a nozzle tip and then material deposition onto a build platform to create an object. Among those techniques, fused deposition modelling (FDM) is widely employed to print RE structures because of their ability to handle neat polymers and metals. In addition, FDM and direct ink writing (DIW) also enable to process composite materials such as PA/hydroxyapatite (HAP) composites and ABS/carbon nanotube (CNT) composites [10]. This study applied FDM to manufacture sandwich structures, consisting of three types of cores, namely conventional honeycomb cells, RE honeycomb cells and tetrachiral cells. Such sandwich structures were fabricated with the building orientation in the longitudinal direction of the structure, and its height was also taken into account on the printing bed. Sandwich structures were manufactured from PLA, which were chosen as the specimens for a three-point bending test according to ASTM C393-20 standard in order to investigate their strength, bending stiffness and energy absorption. The ability of FDM to build RE honeycomb from PU was assessed by Critchley et al. [11]. PU was used as the feedstock material in this study due to its high level of elastic deformation required by cellular structures [39]. Such structures were fabricated along the same direction to prevent layer orientation effect towards mechanical properties of 3D printed parts where the layer height was set to 0.2 mm. On the other hand, arrowhead structures were printed from PU filaments using FDM according to arrowhead design developed by SolidWorks 2015 [31].

Similar to FDM, DIW is also widely used to produce 3D printed parts based on material extrusion. Nonetheless, this technique exclusively employs feedstock materials in ink instead of solid filaments. The ink is ejected out of a nozzle with controlled flow rates and deposited along defined paths onto a build platform to build an object. DIW technique was implemented to manufacture samples originating from alginate-polycaprolactone (PCL)-hydroxyapatite (HA) ink, which is also called APH ink [12]. The ink was prepared at three steps, consisting of the fabrication of PCL/HA/dichloromethane (DCM)/dibutyl phthalate (DBP) mixture, alginate solvent, as well as alginate-PCL-HA (APH) mixture. A main challenge of DIW is to create the ink with a proper viscosity, which can be dispensed out through a nozzle tip, and then rapidly hardened after being deposited onto the build platform.

Powder bed fusion like selective laser sintering (SLS) was employed to fabricate conventional honeycombs [34]. Additionally, the combination between honeycomb and RE honeycomb structures, known as AuxHex structures, was created from nylon powders. It was manifested that AuxHex structures possessed higher tensile strength, specific modulus and specific energy absorption than conventional honeycomb counterparts. Xi et al. [40] also used SLS technique to manufacture a symmetric RE honeycomb (S-RH) structure, as well as a symmetric and rotatable RE honeycomb (AR-RH) structure made from 316L mineral powders. Overall, AR-RH was proven to yield much higher energy absorption capacity when compared with S-RH [40].

Selective laser melting (SLM) implements a heat source to carry out a full melting process for manufacturing a part. SLM was selected to produce RE honeycomb cells based on AlSi10Mg powders in order to fill PU foam [35]. This study compared the energy absorption capacity between empty RE honeycomb (ERH) structures and foam filled RE honeycomb (FFRH) structures. The results indicated that FFRH possessed higher plateau stress and energy absorption capacity than ERH. Besides, the auxetic behaviour of RE honeycomb structures resulted in biaxial compression of the foam, leading to the increase in compression resistance [35].

SLM method has been used to manufacture conventional honeycomb and RE honeycomb structures from 316L metal powders [41]. Initially, a stainless base plate was inserted into the printer cavity. Further, it was filled with 316L powders. The next step was to melt this material at high temperature levels when the sitting head of 3D printer moved to a specific location. Once a layer was created, the printing process continued until the whole part was built. After the printing process was completed, remaining powders clung to the specimen surfaces. As such, the airflow along with vibration treatment was required to avoid the condensation of powders on both internal and external surfaces of 3D printed parts. Besides, those parts also should be heated at 650 °C for 20 h to eliminate residual stress, which were further moved to a closed container under a hot isostatic press in order to minimise the porosity by no more than 1% [42]. Finally, a high-pressure water pistol cut treated specimens from the base plate made of stainless steel. It was reported that RE honeycomb structures possessed higher plateau stress and specific energy absorption with less deformation, as opposed to conventional honeycombs in the same geometry. Moreover, a small side length-to-thickness ratio and input impact energy might improve energy absorption efficiency [41].

Multi jet fusion (MJF) belongs to PFB family by using UV-sensitive photopolymers as feedstock materials in a UV curing system to deposit the droplets in order to harden them properly. This technique was reported to fabricate RE honeycomb, tetrachiral and anti-tetrachiral structures made from PA12 [43]. First of all, a thin layer of material powders was spread onto a build platform. Afterwards, a thermally reactive black ink with fusing agents was sprayed onto a powder bed, in which PA12 particles were expected to fuse together. The detailing agents were then deposited on the part to enhance fusing effect. The entire part was exposed to UV light. This light ignited a thermal reaction between PA12 particles and fusing agents in order to selectively fuse the particles. Non-fused powders were removed from the build platform for future reuse. Once a part layer was made, the build platform moved downwards and another layer of PA12 powders was dispersed on top of it. This process was then repeated until the whole part was fully built. The results indicated that RE-chiral auxetic (RCA) structures generated much greater stress than RE honeycombs, tetrachiral honeycombs and anti-tetrachiral honeycombs. Additionally, RCA structures could be considered as potential energy absorbers under quasi-static uniaxial loading [43].

Electron beam melting (EBM), as part of PBF techniques, depends primarily on an electron beam as the heat source and manufactures a part layer by layer by melting metal powders under high vacuum. This method was implemented to create RE honeycomb structures from Ti6Al4V powders [36] since EBM could produce excellent isotropic properties [44]. It was revealed that small deflection in relation to Timoshenko model induced more adequate accuracy than large deflection model for bending-dominated auxetic structures when subjected to physically realistic stress levels. Hence, the effect of axial extension/contraction owing to bending could be ignored relative to combined shear and bending impacts. Besides, RE auxetic structures possessed insignificant sensitivity to the number of unit cells. The size effect could be negligible when the number of unit cells was more than or equal to 3.

Stereolithography (SLA) among VP methods typically uses a single laser, which is directed on particular spots to cure feedstock materials. Photocurable resin was applied to harden a pattern to create an object. This technique has been well utilised to build RE honeycombs for internal structures of heel pads, and thus, peak plantar pressure of diabetic mellitus (DM) patients could be decreased accordingly [37]. 3D printed heel pads with an auxetic structure yielded a distinct reduction in the peak contact force (~10%) and the mean pressure (~14%), as opposed to a conventional diabetic insole obtained from PU foam. SLA was also employed to manufacture four types of RE structures, namely RE honeycomb (RH) structures, augmented RE honeycomb (ARH) structures, double arrowhead (DAH) structures and augmented double arrowhead (ADAH) structures made from photosensitive resins [3]. All structures were fabricated in a parallel orientation to printing surfaces in order to diminish the anisotropy of bulk materials influenced by printing orientation. A rectangular rigid beam with a width of 25 mm was added to the structure ends to yield more uniform displacement distribution under tensile loading. It was evidently shown that ARH and ADAH structures achieved higher Young's modulus and yield strength when compared with RH and DAH structures [3].

DLP belongs to VP methods to implement photopolymers as feedstock materials and UV light or laser for material solidification. However, this technique employs a projector to direct heat source on specific spots in order to harden the materials rather than directly focus on UV light or laser at the desired points. DLP technique has been applied to produce many structures such as octet-truss, tesseract, star, cross, diamond, vintiles, X-shape, grid, kelvin and RE honeycomb structures using UV-sensitive resins [38]. The results indicated that the grid had a better response when subjected to compressive loading, but instead showed the weakest response to three-point bending. In contrast, star and diamond structures possessed higher stiffness than others so that they might be the best option for typical sandwich structures according to compression and bending mechanisms [38].

Another popular AM method, namely material jetting (MJ) can be used to build conventional honeycomb, RE honeycomb and RE-chiral structures [32]. This technique allows for the fabrication of a part using the specific method similar to a 2D inkjet printer and employs particularly designed ink as feedstock materials. In general, such ink should be heated up to 30–60 °C in order to reach required viscosity [45]. The ink, which is in microdroplets and derived from photopolymer resins, is ejected onto the print bed by the nozzle. The UV curing lamp is employed in MJ to harden deposited droplets. Such a technique commonly requires additional materials to support main structures. Throughout the printing process, the elevator gradually moves down the print bed. It was reported that mutual constraints between auxetic reinforcements and matrices could enhance mechanical properties by obtaining additional support from the matrix phase [32]. Furthermore, the degree of auxetic behaviour might be employed to vary the energy absorption and the stiffness of lattice-reinforced composites.

4.3 Modelling Approaches

Modelling approaches are commonly initiated with topographic analysis and geometric reconstruction for the description of their structure profiles and dimensions. Afterwards, it is followed by analytical modelling using theoretical formulae to define material properties as the responses like compressive strength, Young's modulus, specific energy absorption, as well as Poisson's ratio. Structural profiles along with associated dimensions, governing equations and input parameters, comprising material properties, loading conditions, element size and refinement,

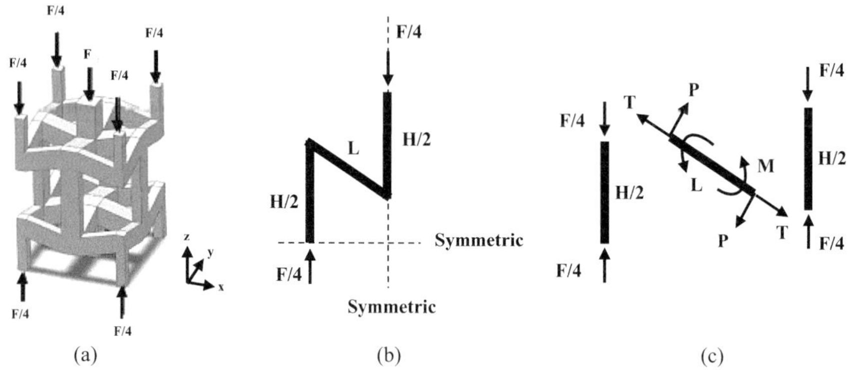

Fig. 4.4 **a** Loading condition in the unit cell of RE structure, **b** specific loads and **c** force components of a 2D simplified structure [36]

are crucial to set up robust finite element models. The last section of modelling approaches is related to the optimisation of material structures, as well as resulting properties in good accordance with experimental work.

An ordinary honeycomb in common hexagonal cells possess in-plane isotropy while a RE honeycomb with RE hexagonal cells is highly anisotropic [14]. Accordingly, it has higher transverse Young's modulus and shear modulus than those of conventional honeycomb structures [15]. Transverse shear modulus depends on a rib slenderness ratio (*w/l*) and approaches the upper limit at a small *w/l* [46]. Meanwhile, a 3D re-entrant honeycomb structure constructed from 2D RE cells showed NPRs in all three principal directions [36]. The symmetry of the unit cell indicated identical mechanical properties along both *x* and *y* axes. Therefore, structural properties were only analysed in two directions, namely *z*-axis and either *x* or *y*-axis, as illustrated in Fig. 4.4.

The effect of density on in-plane Poisson's ratios and Young's moduli of both conventional and RE honeycomb structures were investigated by Whitty et al. [47]. Finite element (FE) models of those structures were in good accordance with experimental results obtained by Alderson et al. [48]. When vertical rib thickness became smaller, the stiffness (E_y) declined accordingly along with smaller NPR (ν_{yx}) [47]. Vertical ribs did not undergo deformation under loading in the *x*-direction, signifying that such ribs were redundant with respect to their mechanical properties in the same direction. Nonetheless, the reduction of diagonal rib thickness induced a decrease in Young's modulus as expected [47]. Conversely, when diagonal rib thickness diminished, in-plane Poisson's ratios of conventional honeycomb and RE honeycomb structures became more negative. A similar result also occurred when vertical and diagonal rib thicknesses were simultaneously reduced [47]. The flexing of diagonal ribs controlled the deformation when their thicknesses were similar to or smaller than that of vertical ribs. Meanwhile, as loaded in the *y*-direction, the stretching of vertical ribs controlled the deformation mechanism when their thickness was significantly lower than that of diagonal ribs [47].

The addition of a narrow rib in RE honeycomb structures significantly increases Young's modulus [49]. Furthermore, a linear relationship between Young's modulus and Poisson's ratio could be achieved by changing additional rib thickness when subjected to small strain tension. On the other hand, Poisson's ratio could shift to positive values when a force constant of additional rib was changed. The effects of RE angle and cell rib length ratio (*h/l*) on the degree of auxeticity and the stiffness of RE honeycombs were investigated by Yang et al. [50]. This study indicated that the degree of auxeticity was independent of RE angle and *h/l*. In contrast, Lee et al. [51] found that the stiffness of RE honeycombs declined with increasing RE angle, resulting in auxeticity enhancement for the structures.

Three assumptions are usually proposed in the unit cell of a RE honeycomb structure for analytical modelling purpose [36]. A unit cell is initially placed in an infinite structure and the infinite number of unit cells is made in all x, y and z directions. As such, boundary effects can be eliminated along with well-retained maximum structural symmetry. Secondly, all joints in RE honeycomb are assumed to be rigid. The deformation of the structure is primarily induced by the bending of RE struts, as well as axial compression of vertical struts. Finally, torsional effect would be generally neglected, and axial shrinkage of RE struts is ignored when normal stress is not taken into account. Overall, the impact on beam deformation is trivial when compared with shearing and bending mechanisms [36]. This study evaluated the influence of uniaxial compression on RE structure in z and y axes due to the symmetry of unit cells along x and y axes.

Figure 4.5 displays a loading analysis of the unit cell in a RE structure when subjected to compressive stress in the z-direction. σ and F denote remote compressive stress and force, while P and T are force components. F, P and T can be calculated using Eqs. (4.1)-(4.3). Meanwhile, Poisson's ratio (ν_{ZX}), effective modulus (E_z) and compressive strength (σ_z) in the z-direction can be determined using Eqs. (4.4)–(4.6) where $\alpha = H/L$ and relevant values of L, t, σ are 5 mm, 1 mm and 10 kPa respectively. In contrast, modulus of elasticity (E) and modulus of rigidity (G) for the used material are 114 and 43 GPa accordingly [36].

$$F = 2\sigma L^2 \sin^2\theta \tag{4.1}$$

$$P = \frac{1}{2}\sigma L^2 \sin^3\theta \tag{4.2}$$

$$T = \frac{1}{2}\sigma L^2 \sin^2\theta \cos\theta \tag{4.3}$$

$$\nu_{ZX} = -\frac{\left(\frac{L^2}{E} + \frac{6t^2}{5G}\right)\cos\theta(\alpha - \cos\theta)}{\frac{L^2\sin^2\theta}{E} + \frac{6t^2\sin^2\theta}{5G} + \frac{4\alpha t^2}{E}} \tag{4.4}$$

$$E_z = \frac{(\alpha - \cos\theta)}{\frac{2\alpha L^2\sin^2\theta}{\mathrm{Et}^2} + \left(\frac{L^4}{2\mathrm{Et}^4} + \frac{3L^2}{5\mathrm{Gt}^2}\right)\sin^4\theta} \tag{4.5}$$

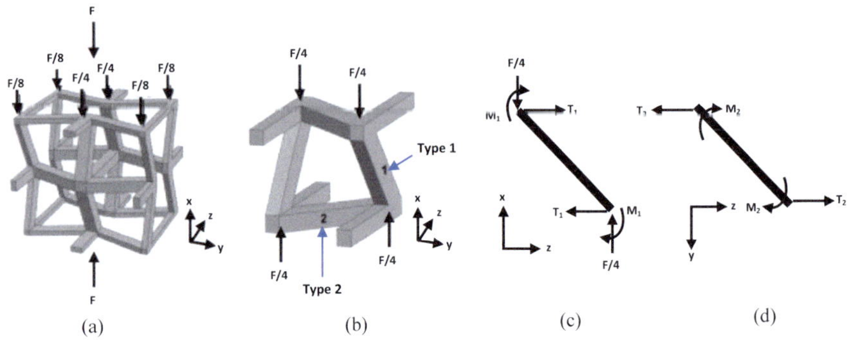

Fig. 4.5 **a** Max-symmetrical unit cell, **b** simplified structure, **c** RE strut type 1 and **d** RE strut type 2 [36]

$$\sigma_z = \frac{F_z}{2L^2 \sin^2 \theta}. \tag{4.6}$$

The similar approach is implemented for modelling uniaxial compression on RE honeycomb structures in y-axis, in which compressive loads of RE honeycomb are illustrated in Fig. 4.5. Force components comprising P and T for strut types 1 and 2 are equal according to geometrical symmetry. Therefore, the formulae to calculate F, P_1, P_2, T_1 and T_2 are presented in Eqs. (4.7)–(4.9). On the other hand, Poisson's ratio (ν_{yz}), effective modulus (E_y) and compressive strength (σ_y) in y-axis can be determined based on Eqs. (4.10)–(4.12).

$$F = 4\sigma(H - L\cos\theta)(L\sin\theta) \tag{4.7}$$

$$P_1 = P_2 = \frac{F}{8}\cos\theta \tag{4.8}$$

$$T_1 = T_2 = \frac{F}{8}\cot\theta \tag{4.9}$$

$$\nu_{yz} = -\frac{\sin^2\theta}{\cos\theta(\alpha - \cos\theta)} \tag{4.10}$$

$$E_y = \frac{1}{L^2\cos^2\theta(\alpha - \cos\theta)} \cdot \frac{1}{\left(\frac{L^2}{\mathrm{Et}^4} + \frac{6}{5\mathrm{Gt}^2}\right)} \tag{4.11}$$

$$\sigma_y = \frac{F_y}{4(H - L\cos\theta)(L\sin\theta)}. \tag{4.12}$$

Equations (4.13)–(4.16) are applied to determine nominal stress σ, nominal strain ε, energy absorption efficiency $\beta(\varepsilon)$ and specific energy absorption E_{SEA} on two distinct types of RE honeycomb structures, which are known as symmetric RE honeycomb (S-RH), as well as asymmetric and rotatable RE honeycomb (AR-RH) [40]. F and A are the compression reaction of the drop hammer and the cross-sectional area of the testing specimen. Meantime, δ, h and M represent the compressive displacement, the length of computer aided design (CAD) model in the vertical direction and the mass of RE honeycomb structure [40].

$$\sigma = \frac{F}{A} \tag{4.13}$$

$$\varepsilon = \frac{\delta}{h} \tag{4.14}$$

$$\beta(\varepsilon) = \frac{\int_0^\varepsilon \sigma(\varepsilon) d\varepsilon}{\sigma(\varepsilon)} \tag{4.15}$$

$$E_{\mathrm{SEA}} = \frac{\int_0^\varepsilon \sigma(\varepsilon) \mathrm{d}\varepsilon}{M}. \tag{4.16}$$

The onset densification strain ($\varepsilon_{\mathrm{cd}}$) along with absorbed energy per unit volume (W_v) are required to calculate specific energy per unit mass SEA. $\varepsilon_{\mathrm{cd}}$ can be determined according to Eq. (4.17) [52]. In contrast, W_v and *SEA* are calculated using Eqs. (4.18) and (4.19)

$$\frac{\mathrm{d}\beta_a(\varepsilon_a)}{\mathrm{d}\varepsilon} |\varepsilon_a = \varepsilon_{\mathrm{cd}} = 0 \tag{4.17}$$

$$W_v = \int_0^{\varepsilon_{\mathrm{cd}}} \sigma(\varepsilon) \mathrm{d}\varepsilon \tag{4.18}$$

$$\mathrm{SEA} = \frac{W_v}{\rho}, \tag{4.19}$$

where σ and ρ denote nominal stress and structure density respectively.

Three types of struts were employed to design 3D RE honeycomb structures, as mentioned earlier [53]. The first type was subjected to axial loading only, while the second and third types underwent flexural and axial loadings. This study implemented analytical, numerical and experimental methods to analyse compressive strength, compressive modulus and PR of RE honeycombs. Meanwhile, Euler–Bernoulli and Timoshenko theories [54] were applied for analytical solutions. Conversely, a numerical method according to volumetric and beam elements was implemented for the validation of analytical results. The results of derived analytical and numerical models were then compared with experimental data. RE honeycomb unit cells were considered to be symmetrical in analytical models. Additionally, total deformation for each degree of freedom (DoF) in the system was assumed as the superposition of separate deformation generated by individual loads of each DoF. Global deformation of cantilever beam was considered as the resultant of four different types of deformation located at the free end of the beam.

An analytical solution has been applied to defined Young's modulus and PR of RE honeycomb [49]. The numerical approach involving finite element (FE) simulations was implemented to validate analytical solution outcomes. This study assumed RE honeycomb unit cell should be symmetrical, signifying that only stretching and flexing applied in the analytical solution. Senobar et al. [55] employed an analytical approach to determine Young's modulus and PR of warp and woof (WAW) structure. This structure was fabricated based on RE honeycomb in three perpendicular directions under uniaxial compression. In this study, a unit cell was selected as a representative volume element (RVE). Furthermore, analytical formulations in the elastic region and lumped plasticity concept were used to determine mechanical properties of WAW unit cells.

Since finite element analysis (FEA) can be adopted to model RE honeycomb structures [36], sample size was considered to be sufficiently large in order to neglect boundary effects for a simplified modelling process. Nonetheless, for a smaller number of unit cells, asymmetrical boundary conditions induced crucial impact on structural performance owing to local stress concentration known as the size effect. The eight repetitions have been recommended by previous study [56] for the unit cell to avoid it. It was estimated that for RE honeycomb structure the size influence was negligible when Poisson's ratio approached -1 [2].

FEA could be carried out to assess the size effect of RE honeycomb structures according to Fig. 4.6a [36]. Those structures were modelled with 3 × 11 unit cell repetitions in two orientations (i.e. width and length) and 1–10 unit cell repetitions in the third direction (i.e. height). On the other hand, boundary conditions for the simulation carried out by SolidWorks COSMOS were revealed in Fig. 4.6a. It was evidently shown that RE honeycomb model was fixed at one end, and normal loading was implemented to the other end. To limit size and boundary effects in the loading direction, this study selected 11 unit cells. Three repetitions on one side of RE honeycomb structure were performed to ensure that the centre layer yielded the maximum geometrical symmetry in order to reduce the size impact at this layer. The displacements at the centre layer were utilised to calculate Poisson's ratio. The effective modulus of the structure was defined by dividing the stress by the total strain of a CAD design. Conversely, the strength of RE honeycomb was achieved with the determination of the minimum stress level when the yield strength of Ti-6Al-4V reached approximately 1050 MPa in any strut [36].

ABAQUS has been used to analyse RE honeycomb structures with a linearly elastic behaviour under tensile loading [49]. RE honeycomb was designed to evaluate in-plane properties of the structure. Tensile loading was imposed in two directions, namely horizontal (1) and vertical (2) directions with the use of boundary conditions depicted in Fig. 4.6b. u_1 denoted a given uniform displacement of right boundary when the load applied in the horizontal direction. In this condition, u_2 represented corresponding displacement at the top boundary due to free deformation of an FE unit cell in the vertical direction. In contrast, when the load in the vertical direction was imposed, u_2 denoted a given uniform displacement at the top boundary in direction 2.

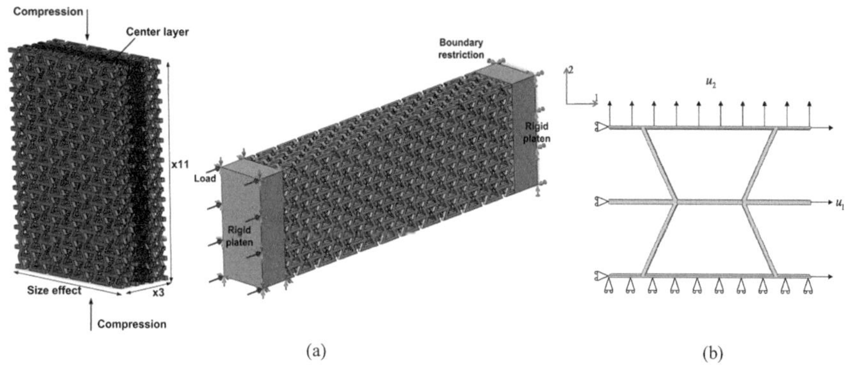

Fig. 4.6 **a** FE models with the size effect and its boundary condition [36] and **b** boundary conditions for RE honeycomb unit cell [49]

Meantime, u_1 was corresponding uniform displacement at the right boundary owing to free deformation of an FE unit cell in the horizontal direction. This study adopted C3D8R quadrilateral eight-node linear brick with the reduced integration to set up numerical models. Additionally, two elements of each unit length were opted for good data convergence.

Designed FE models were also used for RE honeycomb, tetrachiral and anti-tetrachiral structures via ABAQUS [43]. The software was employed to analyse material characteristics under quasi-static compression. This study adopted a loading velocity of 100 mm/s because the modelling process required much more time for a velocity of 0.066 mm/s. Moreover, preliminary simulation showed that FE predicted results were insensitive to a loading velocity below 100 mm/s. Additionally, standard shell elements (S4R) with four-node reduced integration were employed to create structure shell elements. Four levels of element size, consisting of 0.5, 1, 2 and 4 mm, were applied for mesh sensitivity. FE results revealed that element sizes of 0.5 and 1 mm were identical in the elastic regime with the divergence being less than 10% until an axial strain reached 0.34 with valid convergence solutions. Accordingly, an element size of 1 was opted to set up FE models. The general contact algorithm for ABAQUS/Explicit package was implemented to analyse the contact between the specimen and the top and bottom platens along with the interaction between the walls of the structure. The validation of FE models was performed through the comparison of simulated deformation modes, Poisson's ratios and stress–strain curves with experimental data for good agreement.

Mechanical characteristics of auxetic structures subjected to compression was analysed using ABAQUS as well [55]. This study applied WAW made from RE honeycomb. The raw material was assumed to possess a perfect elastic–plastic material behaviour. Furthermore, quasi-static analysis was employed in the simulations in order to overcome convergence issues in the plastic region. On the contrary, Newton's incremental algorithm and smooth loading method were selected to reduce analysis time and solve nonlinear formulae. The minimum increment size could reach 10^{-7}, and the convergence was based on the criterion that residual errors must be less than 0.5%.

ANSYS/LS-DYNA was employed to design symmetric RE honeycomb (S-RH), as well as asymmetric and rotatable RE honeycomb (AR-RH) structures in FEA [40]. Such designs were created with shell element 163 and full integration of Belytscho-Tsay shell element algorithms to increase the convergence. S RH and AR-RH designs were located between two rigid plates, and each cell surface was considered as a single self-contact surface. In addition, surface-to-surface contact applied between the specimen and two rigid plates. On the other hand, SR4 elements were adopted for model mesh, while applied mesh method was swept using quad-dominated elements.

ANSYS was also used to model RE honeycomb structures and calculate their mechanical properties based on beam and volumetric elements [53]. This study applied Euler–Bernoulli and Timoshenko beam theories in an analytical approach. Transverse shear deformation was taken into account in this study, and all struts were considered to be rigidly connected to each other at the vertices. Besides, linearly elastic characteristic data of selected materials achieved from mechanical tests was employed in FE models. Meanwhile, overlapping impact at the joint on mechanical properties of RE honeycomb was assumed in volumetric elements. Furthermore, adaptive fine volumetric mesh was set up for discretising the models according to tetrahedral 10-node elements to yield fine mesh at the vertices, as well as coherently larger mesh at the middle of the struts.

Generally speaking, simulation analysis and optimisation algorithm are used to be integrated with experimental outcomes in order to yield optimum RE structures with desired properties [57]. Regarding this, multi-objective optimisation enables the combination of both aspects [58]. Meantime, non-dominated sorting genetic algorithm (NSGA)-II was adopted in [59] to optimise design parameters of crash boxes filled with lattice structures, consisting of conventional honeycomb, RE honeycomb, triangular and Kagome structures. NSGA-II was classified as an evolutionary algorithm, which was created because of the deficiency of previous evolutionary algorithms in a lack of elitism, and applied a sharing parameter to maintain a diverse Pareto set [60]. Ma et al. [61] carried out an optimal design of crash boxes according to response surface methodology (RSM) and multi-island genetic algorithm (MIGA). This study focused on structural design of collision boxes, as well as optimal improvement of crashworthiness performance. In general, RSM can be used to build the relationship between explanatory variables and response variables. This method relies on a sequence of designed experiments to gain the best response that could be either the maximum, the minimum or a specific target. Choudhry et al. [62] also adopted RSM to calculate maximum and minimum energy absorptions of RE honeycombs. This method appears to be appropriate to evaluate the relationship between geometrical parameters on the responses in order to explore mechanical characteristics of auxetic nasopharyngeal swab structures [63]. Etemadi et al. [64] also used a similar method for conducting parametric analysis by comparing the significance of each design parameter on the responses. Hence, the most and least significant geometric parameters to elastic constants could be determined accordingly.

Central composite design (CCD) approach according to a five-level design of experiments (DoEs) and FEA-based factorial study with associated optimisation were employed to investigate the influence of geometric parameters on PR of RE honeycomb structures [65]. This study involved RE angle θ, length of vertical strut H and length of inclined strut L as geometric parameters. The output parameter of DoEs was the optimum factor-level combination, resulting in the maximum NPR using RSM. It was reported that the length of inclined strut and RE angle induced greater influence on transverse displacement and Poisson's ratio despite the minimum effect relating to the length of vertical strut. A better auxetic characteristic was achieved at a lower θ, while increasing the number of unit cells from 1 to 4 declined Poisson's ratio. It could be associated with the enhanced RE structure stiffness owing to the increase in the number of unit cells, thus resulting in reduced transverse displacement [65].

Carakapurwa and Santosa [66] employed FEA to determine specific energy absorption (SEA) of RE structures. This study created four different lattice structures, comprising RE honeycomb structures, double arrowhead structures, star-shaped structures, as well as double-U structures. The associated parameters for near-random samples, originating from a distribution of multidimensions, were designed based on Latin hypercube sampling (LHS) method. Meanwhile, machine learning method was implemented to determine the design with the maximum SEA. The optimisation in this method comprised two steps, namely model approximation using artificial neural network (ANN), as well as variable optimisation obtained from NSGA-II. The optimisation outcomes suggested that star-shaped and aluminium structures with the thickness of 2.95 mm possessed the maximum SEA, which was 1220% higher than that using a baseline model [66].

The issues related to the limited amounts of material distribution in a design space can be overcome by topology optimisation (TO) method [68]. Solid isotropic material along with the penalisation (SIMP), as well as the method of moving asymptotes (MMA) to simulate TO for RE honeycomb structures was implemented, as mentioned earlier by Strek et al. [67]. Besides, SIMP and MMA were also applied to demonstrate that the spread-out of soft and hard materials in a two-phase RE honeycomb could alter mechanical performance of RE honeycomb, along with the utilisation of FE method in structural analysis. The geometry of quarter RE honeycomb with associated boundary conditions, the distribution of optimised constituents, as well as deformed shape after compression were depicted in Fig. 4.7. t, L and α represented rib thickness, rib length and angle of rib inclination, respectively. On the other hand, G_1 and G_2 denoted the boundaries on the top and side surfaces of RE honeycomb. Once the structure was compressed, boundary G_2 was moved down. Meantime, a structure domain contained 20–40% of hard materials. It could be seen that two-phase RE honeycomb might change its PR from negative to positive values, which was influenced by its material distribution and mechanical properties, as evidenced by Young's modulus and PR of the structure. The mechanical behaviour of two constituents greatly influenced resulting advanced structures.

Taherkhani et al. [69] implemented TO method to optimise a material distribution of RE honeycomb. This study was based on modified solid isotropic materials

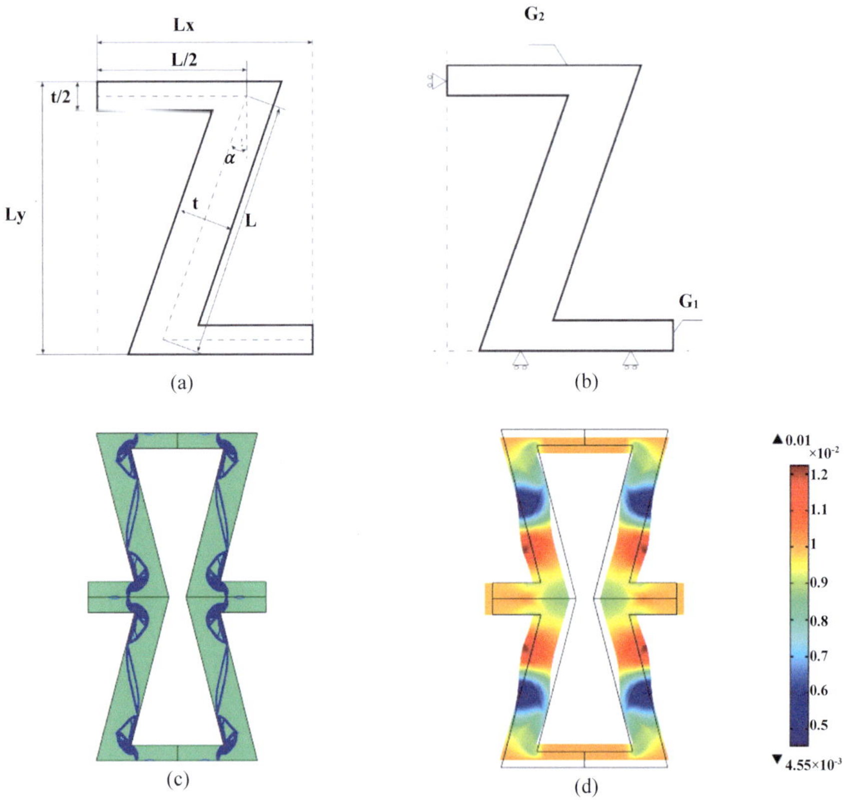

Fig. 4.7 **a** Geometry of the quarter, **b** boundary conditions, **c** optimised constituent distribution (hard material–blue, soft material–green) and **d** deformed shape of a RE honeycomb model [67]. Colour map represents the value of total displacement (m)

and the penalisation TO method for RE honeycomb structures with different PRs and volume fractions. The structures created were then numerically designed by FE method and further validated using experimental tests. The related data indicated that PRs were affected by the location and stiffness of rotation joints. PRs decreased when the position of rotation joints was closer. Besides, the reduction in the stiffness of rotation joints also declined PR values for the structure. Kaminakis and Stavroulakis [70] adopted TO method to model auxetic microstructures of materials that involved the combination of FE method and evolutionary-hybrid algorithms. Meanwhile, Lukasiak et al. [71] proposed a new algorithm of numerical inverse homogenisation for planar isotropic composites according to periodic hexagonal cells. The internal structure of a hexagonal cell is considered to have an angle of 20° for rotational symmetry. It yields a notable depletion in the number of design variables to the optimised analytical problem, which is approximately 6 times less than that in a rectangular shape.

The main issue in TO might be associated with the uncertainty when iterative efficiency significantly affects the results. It was reported that a highly dense FE mesh (10^3) was used in the optimisation of 3D material microstructures along with an auxetic behaviour for numerical accuracy [72]. However, a large number of iterations up to 3000 limited widespread applications of conventional TO methods, which was employed to identify microstructures of novel materials. To a great extent, geometric symmetry may be applicable to material microstructures in order to restrict design freedom [73]. Nonetheless, minimised design space could diminish the possibility of developing novel auxetic microstructures. As such, a more effective TO method is required to model 3D auxetic metamaterials that might be applicable to auxetic structures.

4.4 Multifunctional Properties

Re-entrant structures have unique constructions that distinguish them from other solid structures. It makes those structures possess particular material characteristics like high NPR [43], exceptional tensile and yield strengths [34], extraordinary energy absorption [10], remarkable shape memory effect and strain recovery [74], as well as excellent electrical conductivity when coated with graphene [75]. This section focuses on mechanical characteristics, shape memory effect and other key properties of RE structures.

Mechanical characteristics of RE structures consist of PR, compressive properties, as well as energy absorption. PR shows material deformation perpendicular to the loading direction. It can be generally formulated as the negative ratio of lateral strain to axial strain. The Poisson's ratio of RE honeycomb, tetrachiral and anti-tetrachiral structures can be determined using Eqs. (4.20)–(4.22) [43]. First of all, nominal strain ε can be calculated by Eq. (4.20) where δ and V represent the changes in displacement between machine platens and original height of the specimen. Afterwards, it can be implemented to define $\varepsilon_{\text{true}}$ using Eq. (4.21). Meanwhile, Eq. (4.22) is applicable to determine Poisson's ratio (υ_{xy}) for all 3D printed structures.

$$\varepsilon = \frac{\delta}{V} \tag{4.20}$$

$$\varepsilon_{\text{true}} = \ln\left(1 + \varepsilon_{\text{engineering}}\right) \tag{4.21}$$

$$\upsilon_{\text{xy}} = -\frac{\varepsilon_{\text{true lateral}}}{\varepsilon_{\text{true axial}}}. \tag{4.22}$$

In relation to the compression in the x-direction, simulated PRs and experimental data were approximately –0.7. It was then gradually declined with reduced engineering strain to –0.8, as demonstrated in Fig. 4.8a. In contrast, PR being determined under compressive loading in the y-direction could reach –1, and then was elevated up to –0.7 when the engineering strain reached 0.12 shown in Fig. 4.8b. *x* and *y* loading orientations refer to *x* and *y* axes according to Fig. 4.8. The auxetic characteristic of tetrachiral structures was induced by cylinder rotation. It was reported by Prall and Lakes [22] that theoretical PR of chiral honeycombs could reach –1, which was lower than corresponding value of –0.78 obtained by Alomarah et al. [43], as observed in Fig. 4.8c. This result might be associated with relatively small lateral deformation. Prall and Lakes [22] did not involve axial deformation and shear deformation of the walls when calculating theoretical PR as a common drawback. On the other hand, PR value of anti-tetrachiral structures was found to be nearly -1, which became slightly declined when increasing engineering strain according to Fig. 4.8d.

Alomarah et al. [43] calculated SEA of honeycomb, tetrachiral, as well as anti-tetrachiral structures based on both FEA and experimental work. FEA predicted results typically could overestimate experimental data of absorbed energy, which arose from the presence of microvoids within 3D printed structures to deteriorate the ability of energy absorption. Furthermore, the loading direction slightly

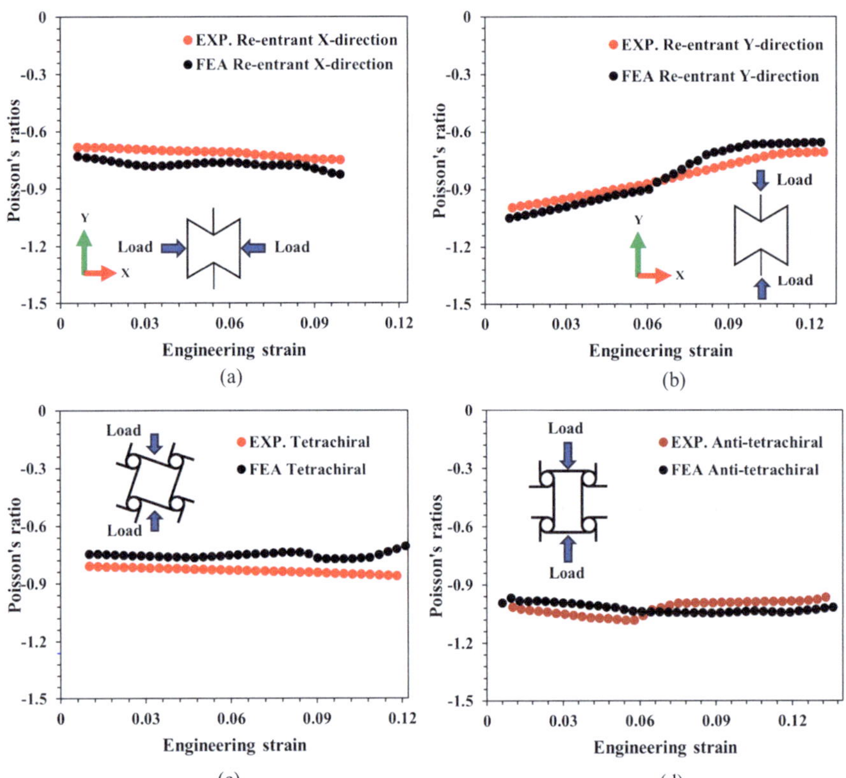

Fig. 4.8 PR curves obtained from experimental and simulated work for RE honeycomb subjected to compressive loading in **a** *x*-direction, **b** *y*-direction, as well as with **c** tetrachiral structures and **d** anti-tetrachiral structures [9]

affected absorbed energy of RE honeycomb where SEA in the x-direction possessed a higher value when compared with that in the y-direction. Concerning this, tetra-chiral structures yielded the highest absorbed energy due to its larger plateau stress when compared with other types [43].

Chow et al. [31] investigated the effect of internal angle on PR in both transverse and longitudinal directions using experimental and simulation methods for double arrowhead structures. In the most cases, it was in good agreement between experimental data and simulation results with increasing internal angle so that it could enhance PR in all loading directions. Arrowhead structures with an internal angle of 30° induced the greatest auxetic impact on PR at the axial strain up to 0.5. Meanwhile, in the longitudinal direction, arrowhead structure possessed a smaller PR than that in the transverse direction. It reflected excellent shape formability of the structures owing to their capability of volumetric expansion under tensile loading. The internal angle of RE honeycomb structures could vary in range of 30°-70° [76]. It was detected that increasing internal angle by 60° enhanced NPR value up to –0.4. Nonetheless, a further increase in internal angle diminished NPR. The higher NPR for RE honeycombs was reached according to Li et al. [77]. This study calculated PR values using FEA for four internal angles consisting of 30, 45, 60 and 75° respectively. It was found that increasing internal angle promoted an auxetic behaviour. The highest NPR of –1.9 was achieved at an internal angle of 75°. It was revealed that larger internal angle enhanced the auxeticity of RE honeycombs.

The influence of relative density on PR and elastic modulus was manifested when using butterfly-shaped auxetic structures [78]. Such structures were deemed the development outcome of honeycombs. It is well known that increasing relative density of butterfly-shaped honeycombs could undermine NPR behaviour despite their higher elastic modulus. Such results were in good accordance with Lan et al. [79] based on designed arrowhead structures to evaluate their damage characteristic under quasi-static uniaxial compression at room temperature. Once the relative density of those structures, indicated by the ratio of strut cross-sectional thickness t, as well as horizontal projection length of oblique strut l, was increased from 0.1 to 0.225, the compressive strength and corresponding SEA were enhanced up to 8 and 7.5 times accordingly. Indres et al. [10] reported that the sandwich structure made from 90° RE honeycomb core with a relative density of 0.25 yielded the greatest energy absorption of 6.58 J. On the contrary, the same structure constructed from 0° RE honeycomb core along with a relative density of 0.15 induced the minimum energy absorption instead. It signified that the density level and orientation angle of the cells significantly affected the ability of structures for energy absorption. It was clearly observed that the relative density of the structure played an important role in its mechanical properties.

RE honeycombs were compared with 8 other different structures, comprising octet-truss, tesseract, star, cross, diamond, vintiles, X-shape and grid structures, as depicted in Figs. 4.9a–i, to determine their compressive strengths [38]. It was revealed that the compressive load of diamond structure reached approximately 420 N, which was considered to be equivalent to the highest compressive strength according to

Fig. 4.9j. On the contrary, the lowest load of 160 N was achieved for RE honeycomb structures. Furthermore, the largest deflection at failure occurred at 4.2 mm for octet-truss structures, as evidently shown in Fig. 4.9k. Nevertheless, a constant-stress plateau could not be identified due to brittle nature of all lattice cores. It was found that several lattice structures, such as RE honeycomb structures, exhibited excellent mechanical behaviour similar to stretch-dominated structures even though the smallest failure load occurred in grid and honeycomb structures [80].

The influence of orientation angle and relative density on bending behaviour of sandwich beams made from honeycomb, RE honeycomb, as well as chiral cells manufactured by FDM, was investigated by Indres et al. [10]. This study used two different orientation angles of RE honeycomb including 0° and 90°. The sandwich beams had relative densities of 0.15 and 0.25 respectively. The greatest bending load per unit length was achieved at 266 N/mm for the sandwich beam built from honeycomb core with a relative density of 0.25. In contrast, a similar structure containing a RE honeycomb core reached the lowest bending load.

RE structures, along with their specific deformation behaviour and porosity, enable them to absorb energy considerably. The impact region of such structures undergoes common contraction under in-plane impact loading. It further improved the stiffness in the impact region, thus increasing their energy absorption [19].

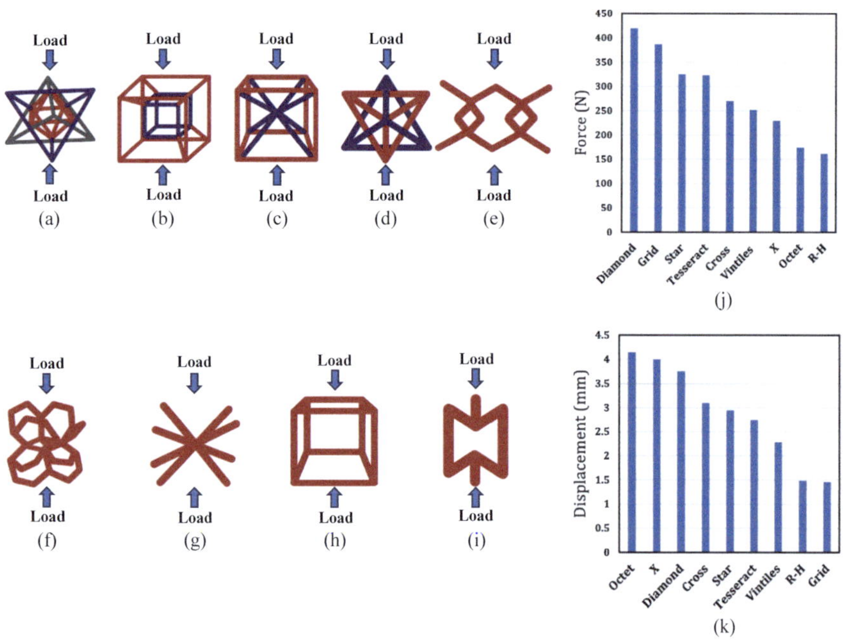

Fig. 4.9 Unit cells with nine different structures: **a** Octet-truss **b** tesseract, **c** star, **d** cross, **e** diamond, **f** vintiles, **g** X, **h** grid, **i** RE honeycomb, as well as **j** force and **k** displacement of lattice core in sandwich structures under compressive loading [38]

Xi et al. [40] evaluated the effect of three different impact velocities of 3, 4 and 5 m/s on absorption capability, as well as deformation patterns with respect to symmetric RE honeycomb (S-RH), as well as symmetric and rotatable RE honeycomb (AR-RH) structures. S-RH and AR-RH structures had similar energy absorptions with their corresponding strain being less than 0.1. At this strain level, the structure underwent elastic deformation, and both structures yielded similar stiffness at the same thickness. However, AR-RH structure provided better energy absorption when compared with S-RH counterpart in the strain range of 0.1–0.4 where local cells of S-RH structures encountered a buckling phenomenon with the growth of collapse stress. Buckling impact can be avoided in AR-RH structures due to their asymmetry and rotatability along with more steady platform stress.

Some distinct unit cells of auxetic structures can be merged to create hybrid cellular (HC) structures, which are essential to obtain desired mechanical properties. Xu et al. [34] merged honeycomb with RE honeycomb structures with different thicknesses to produce HC structures known as AuxHex according to Fig. 4.10a. This combination enabled to enhance compressive strength, SEA, as well as specific modulus of the structures. Figure 4.10b showed that AuxHex structures resulted in higher compressive strength than honeycomb counterparts in both x and y directions. A similar tendency for SEA and compressive modulus was also depicted in Fig. 4.10c and d respectively.

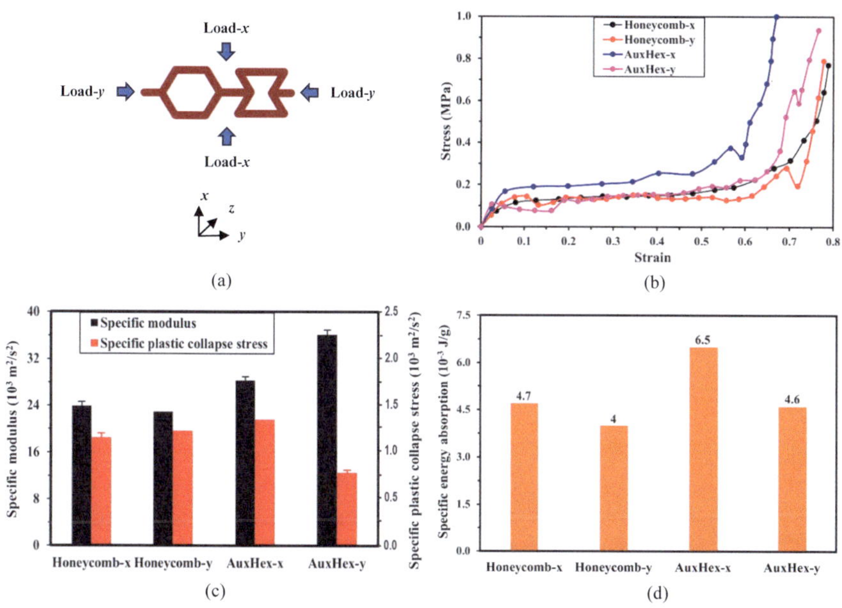

Fig. 4.10 **a** AuxHex unit cell and **b** stress–strain, **c** specific modulus, as well as **d** SEA diagrams for honeycomb and AuxHex structures subjected to compressive loading in x and y directions [34]

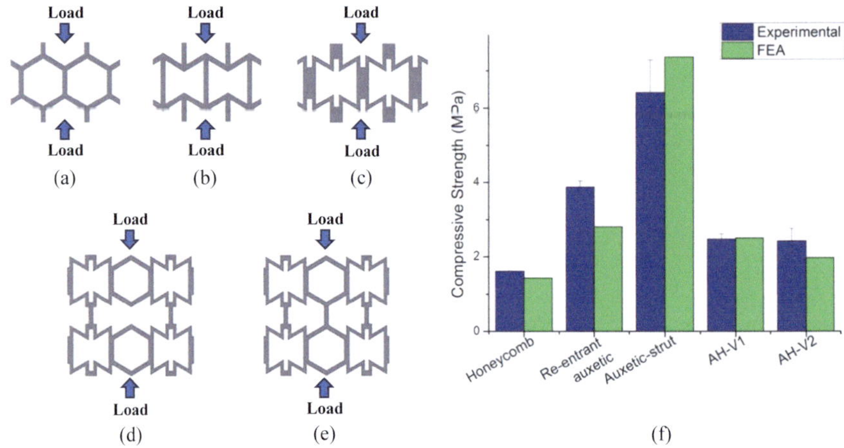

Fig. 4.11 Five typical unit cells: **a** Conventional honeycomb, **b** RE honeycomb, **c** auxetics strut, **d** AH-V1 and **e** AH-V2, as well as **f** compressive strengths of all corresponding structures [81]

Ingrole et al. [81] evaluated compressive strength of another HC structure called auxetic-honeycomb (AH), which was made from conventional honeycomb structures. This study developed RE honeycomb structures to be auxetic struts. On the other hand, AH was built from honeycomb and auxetic struts. Five distinct structures consisting of conventional honeycomb, RE honeycomb, auxetic strut, auxetic-honeycomb 1 (AH-V1) and auxetics-honeycomb 2 (AH-V2) structures are illustrated in Figs. 4.11a–e. This study relied on compressive tests and FEA to determine compressive strengths of such structures both numerically and experimentally shown in Fig. 4.11f. It was clearly seen that auxetic struts exhibited the highest compressive strength, which might be generated by two factors, comprising different deformation mechanisms of cell structures, as well as a critical role of relative density in those structures.

Another important mechanical property is well known as fatigue property, which is necessary for particular applications like aviation, automobiles and orthopaedics, especially with respect to metal components. Choi et al. [82] investigated fracture toughness of copper foams with both positive PR and NPR values. It was reported that the foam with NPR values yielded greater fracture toughness when compared with those with positive PRs. Kramberger [83] studied the influence of unit cell distribution, along with the shape on the fracture characteristics of 2D auxetic structures under quasi-static loading. Relevant data indicated that cell geometry and unit cell distribution significantly affected fracture characteristic of auxetic structures. The effects of radius in the intercellular connections and NPR on the fatigue behaviour of auxetic structures were also mentioned earlier [84]. This study employed a computational analysis technique in order to optimise auxetic cellular structures made from AlSi10Mg-aluminium alloys. It was manifested that the radius in the intercellular connections and NPR significantly influenced the fatigue limit of auxetic structures, in which a longer fatigue limit was acquired by auxetic structures with a higher NPR. The static and ultra-low cycle fatigue (ULCF) properties of thin cellular structures, possessing auxetic characteristics derived from AA 7075-T651-aluminium alloys, were identified in previous studies [85, 86]. Both continuum damage mechanics (CDM) and fracture mechanics approach (FMA) were implemented to evaluate the fatigue characteristic of such structures. It was reported that the energy-based technique used in this study met the expectations to precisely calculate the durability of auxetic cellular structures.

Shape memory effect shows the ability of a material to memorise and regain its initial form after being deformed into a temporary shape under a stimulus. Some materials enable to exhibit such a unique characteristic due to the presence of elastic and transition segments [87]. The shape memory effect of a material can be indicated through its strain fixity, strain recovery, as well as temperature memory effect (TME). The strain fixity (SF) denotes material capability to fix mechanical deformation induced throughout a programming process. It is shown by the deformation percentage, which is fixed in the specimen after load removal. Meanwhile, strain recovery (*SR*) reflects material recovery ability, which can be determined via the deformation percentage being achieved at the end of the heating ramp.

Temperature-dependent memory effect indicates the capability of a shape memory material, especially special shape memory polymers, to remember the shape and the temperature under deformation [88, 89]. This behaviour was evidently performed by amorphous polymers [90] and semicrystalline polymers [91] with either melting or glass transition distributed in a wide temperature range. Meanwhile, temperature-dependent effect for RE honeycombs can be gained through shape memory activation over a thermal ramp called thermally stimulated recovery (TSR) tests when subjected to tension to determine their recovery rate (RR_T) [74]. Nonetheless, prior to the determination of RR_T, all formed changes of the specimens, known as normalised displacement (Δl_{norm}), as a function of temperature, are required to be calculated accordingly. SF, SR, Δl_{norm} and RR_T can be then determined using Eqs. (4.23)–(4.26) as follows:

$$\mathrm{SF}(\%) = \left(\frac{\Delta l_{\mathrm{unload}}}{\Delta l_{\mathrm{appl}}}\right) \times 100 \tag{4.23}$$

$$\mathrm{SR}(\%) = \left(1 - \frac{\Delta l_{\mathrm{final}}}{\Delta l_{\mathrm{appl}}}\right) \times 100 \tag{4.24}$$

$$\Delta l_{\mathrm{norm}}(\%) = \left(\frac{\Delta l}{\Delta l_{\mathrm{appl}}}\right) \times 100 \tag{4.25}$$

$$\mathrm{RR}_T\left(\frac{\%}{^\circ\mathrm{C}}\right) = \frac{\mathrm{d}(\Delta l_{\mathrm{norm}})}{\mathrm{d}T}, \tag{4.26}$$

where $\Delta l_{unload} = l_{unload}—l_0$ and $\Delta l_{appl} = l_{appl}—l_0$. l_{unload} and l_0 represent the clamp distance at load removal and initial clamp distance respectively. Meanwhile, l_{appl}

denotes the clamp distance corresponding to the resulting elongation. $\Delta l_{\text{final}} = l_{\text{final}}$—$l_0$ where l_{final} reflects the clamp distance at the end of the test. On the other hand, $\Delta l = l$—l_0 where l represents the distance between the clamps throughout the tests.

Pasini et al. [74] proposed two programming steps, namely single step and double step using a dynamic mechanical analyser (DMA). In the initial stage of double-step programming process, three different magnitudes of tensile elongation (Δ_{lappl}), comprising 25%, 50% and 75% of total used displacement, were employed at the deformation temperature (T_{def}) of 100 °C, as indicated by 100 (25%), 100 (50%) and 100 (75%) in Table 4.1. On the contrary, the remaining magnitudes of deformation with 75%, 50% and 25% of total applied displacement were applied in the second step at 40 °C for T_{def}, which are referred to as 40 (75%), 40 (50%) and 40 (25%) accordingly.

The results of TSR tests indicated that all specimens yielded excellent strain fixity, as shown by their *SF* values in a range of 65–99%, as well as 84–92% for single-step and double-step programming processes respectively (Table 4.1). When T_{def} and peak temperature (T_{peak}) were increased, SF values mainly for T_{def} above 60 °C or at least above its T_g was also enhanced. Furthermore, the specimens demonstrated good shape memory response by reaching their shape recovery between 95–100% and 94–100% for single-step and double-step programming processes accordingly. *SR* could not be affected by T_{def} when compared with SF. All specimens almost retrieved their original form, which was independent of deformation temperature [74].

The recovery rate describes the process distribution via the temperature derivative of normalised displacement according to Eq. (4.26). In the single-step programming process, lower T_{def} resulted in smaller RR_T with a broader data distribution, as compared to the higher counterpart. It signified that the specimens deformed at lower T_{def} required longer time to regain their original form. The highest RR_T at

Table 4.1 Summary of strain fixity and recovery in TSR tests [74]

TSR test	Deformation temperature T_{def} (°C)	Strain fixity SF (%)	Strain recovery SR (%)	Peak temperature T_{peak} (°C)
Single-step programming	20	63	100	25
	40	77	97	47
	60	90	99	64
	80	98	98	81
	100	99	95	85
Double-step programming	100 (25%)-40 (75%)	84	100	50.85
	100 (50%)-40 (50%)	88	94	45.85
	100 (75%)-40 (25%)	92	94	47.90

5%/ °C might be reached at the deformation temperature of approximately 80 °C. Meanwhile, in the double-step programming process, the samples with the first lower tensile elongation ($\Delta l_{\text{appl, 100 °C}}$) demonstrated a greater recovery rate, but decreased RR_T in additional tensile elongation ($\Delta l_{\text{appl, 40 °C}}$). The recovery process of those samples at the first heating step was much faster when compared with its counterpart for the second heating. Increasing $\Delta l_{\text{appl, 100 °C}}$ could decline RR_T for the first heating step, but improved the recovery rate for the second heating [74].

RE structures possess other common properties like electrical conductivity and piezoresistance when primary materials are embedded with additives like graphene and carbon fibres. The electrical conductivity of RE structures is reflected by their resistance change and piezoresistance. The electrical conductivities of 3D printed auxetic lattice structures, made from four different cell types, namely trust (TR), honeycomb (HN), chiral truss (CT) and RE honeycomb (RE) structures, were investigated in previous work [75]. The resistance change in this study was based on the movement of a graphene network. Once a sample was stretched, graphene particles were split up, making them more distant along with the resistance increase. Electrical conductivities of such lattice structures before and after coating in both counter direction (CD) and machine direction (MD) were displayed in Figs. 4.12 and 4.13. At least four cycles of the coating were required by the electrical current to flow via entire lattice structures because adequate network between graphene and electrical conductivity was not constructed at the first coating.

All structures excluding CT indicated remarkable electrical conductivities in the CD, as depicted in Fig. 4.12, which signified that CT structures possessed adverse paths through which the current flowed due to their complex structures. In the MD, the current conversely showed a smooth flow in TR and HN structures according to Fig. 4.13. It revealed that the current could flow in those structures without the hindrance regardless of different directions. However, the gradients of I-V curves in the MD were lower than those in the CD. Accordingly, it was recommended that this unique behaviour should be taken into account when using RE as a strain sensor.

Andrew et al. [92] investigated the piezoresistance of three lattice structures built from hexagonal, chiral and RE honeycomb cells. This study employed polyether ether ketone (PEEK) and carbon fibres (CFs) as the matrices and the reinforcements in composite systems respectively. Figure 4.14 described a piezoresistive response of PEEK/CF composite lattice structures constructed from all cell types under both in-plane and out-of-plane quasi-static compressions.

Figure 4.14a suggested that the measurement data of normalised resistance change, known as $-\Delta R/R_0$, was enhanced monotonically when the strain for such lattice structures was increased. ΔR represents the measured change in resistance with $\Delta R = R - R_0$, while R_0 indicates initial non-loading resistance. It was evidently seen that relatively brittle PEEK/CF composites gave rise to cell collapse, along with compressive strain and crush band formation. At this level, new connections between struts of lattice structures were constructed with crush bands percolation in the whole specimen. The percolated bands reflect the pathways of low resistance as the origin of decreased resistance [93].

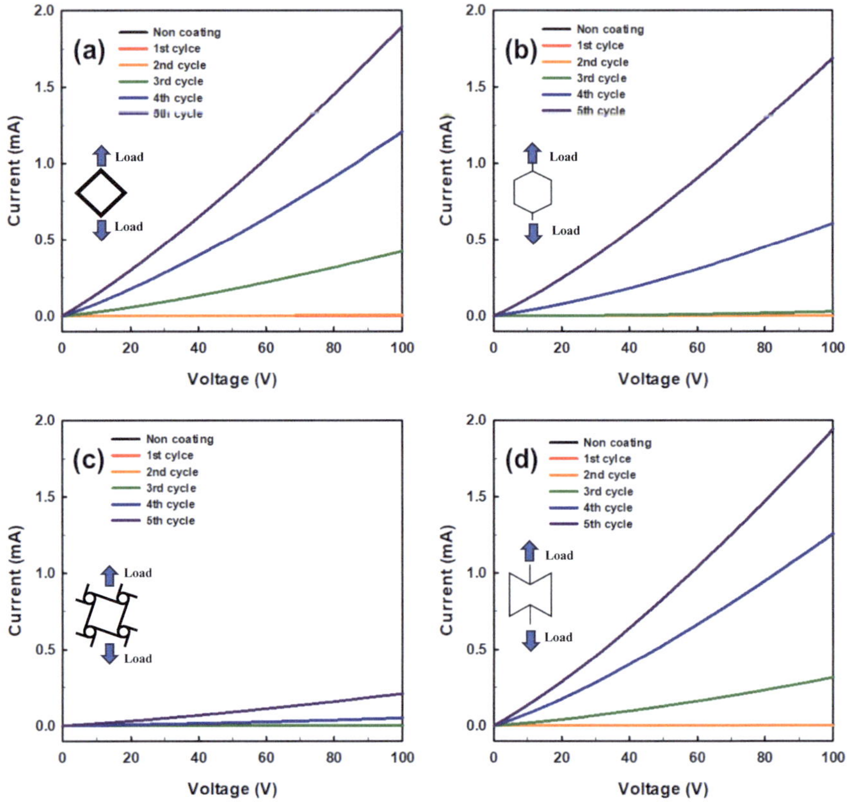

Fig. 4.12 Current–voltage (I-V) curves for non-coating and after coating cycle in CD for **a** TR, **b** HN, **c** CT and **d** RE structures [75]

When the compressive strain was increased, more percolated crush bands could be generated, resulting in nearly complete electrical resistance loss of crushed lattices. Accordingly, nonlinear piezoresistive responses at higher stains were manifested. Figure 4.14b illustrated the curves indicating a relationship between measured $\Delta R/R_0$ and strain response for all lattice types under quasi-static and out-of-plane compression. The piezoresistive response of such structures was weaker than those subjected to in-plane compression. In the area of small strain, $\Delta R/R_0$ values were greater with increasing compressive strain. Conversely, such values underwent a decreasing trend in the area of larger strain, which could be induced by a combination of the decrease in the cell wall subjected to compression, as well as the formation of new contacts because of those folds created among cell walls. The first increase in electrical resistance was related to lateral expansion of cell walls in the elastic area of the responses due to partial damage of conductive path in relation to Poisson's impact.

AlMahri et al. [94] also evaluated the ability of piezoresistive sensing for lattice structures. All lattice structures were manufactured from composite materials using

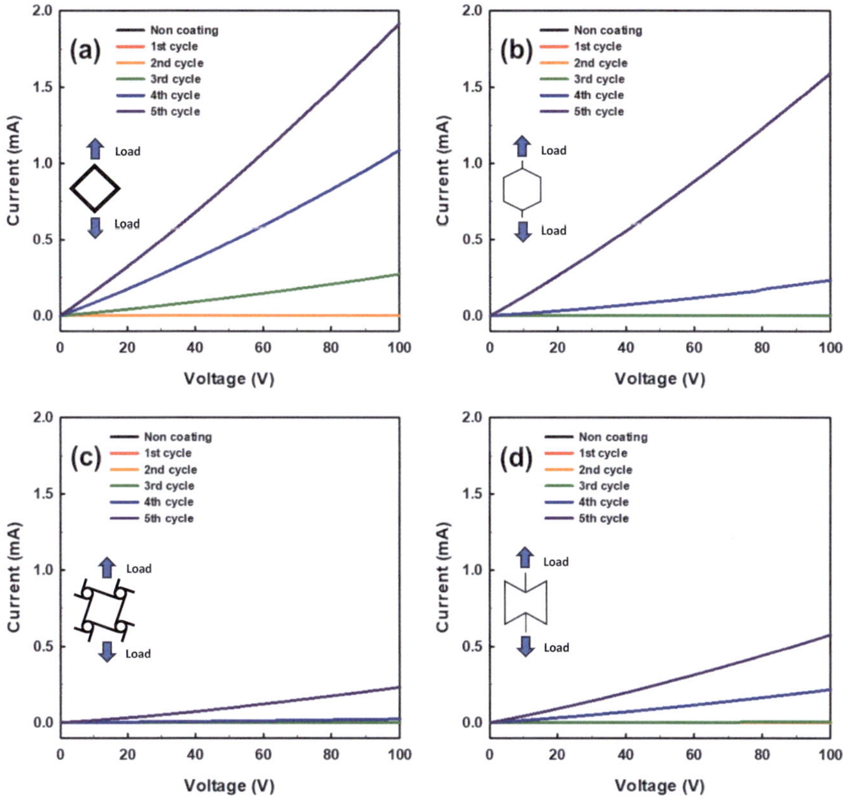

Fig. 4.13 Current–voltage (I-V) curves for non-coating and after coating cycle in MD for **a** TR, **b** HN, **c** CT and **d** RE [75]

high-density polyethylene (HDPE) and multi-walled carbon nanotubes (MWCNTs) as a primary material and additives respectively. Three cell types, consisting of S-shaped, chiral and RE honeycomb structures, were employed to fabricate lattice structures via FDM. Gauge factor k, as formulated in Eq. (4.27), is a critical parameter to calculate piezoresistive responses for those lattice types

$$k = \frac{(\Delta R_2 - \Delta R_1)/R_0}{\varepsilon_1 - \varepsilon_2}, \tag{4.27}$$

where $\Delta R_1/R_0$ and $\Delta R_2/R_0$ denote the resistance changes relative to R_0, which are determined at two applied strains ε_1 and ε_2. The results clearly indicated that S-shaped cellular structures reinforced with 4 wt% MWCNTs yielded $k = 7.6$, which was approximately threefold greater than those based on both chiral and RE honeycomb structures [94].

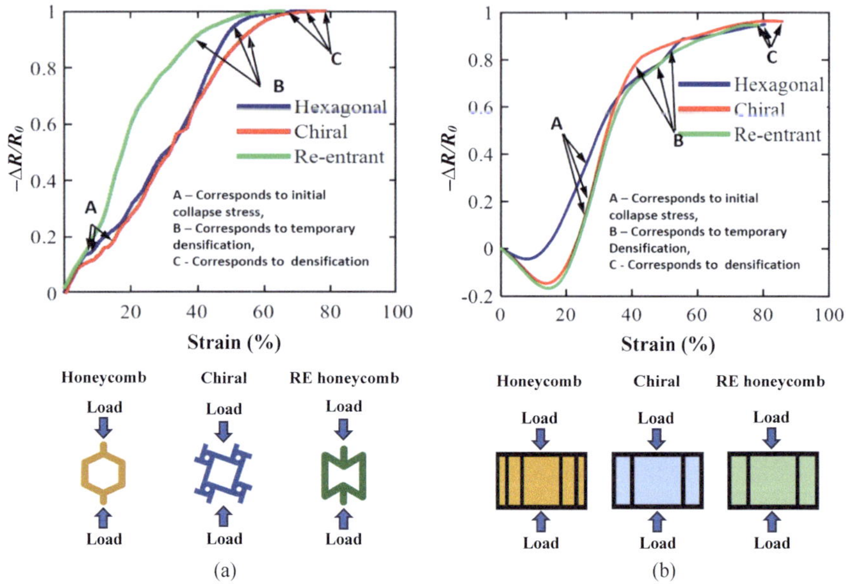

Fig. 4.14 Normalised change in resistance vs strain curves of PEEK/CF composites lattice structures with hexagonal, chiral and RE honeycomb cell topologies subjected to **a** in-plane compression and **b** out-of-plane compression [92]

4.5 Summary

This chapter elaborated four structural designs of RE structures, comprising conventional honeycomb, RE honeycomb, tetrachiral and arrowhead structures, which are typically used as reference models. Due to structural complexity, RE structures are quite difficult to make using conventional manufacturing methods like melt extrusion and injection moulding. The presence of additive manufacturing techniques such as material extrusion, material jetting, powder bed fusion and vat photopolymerisation allows such structures to be fabricated by depositing feedstock materials in a layer-by-layer manner onto the build platform. Topographic analysis, geometric reconstruction, analytical modelling, optimisation of material structures and resulting properties are commonly used in modelling approaches in order to yield the best model with the desired properties. The properties usually investigated in RE structures are Poisson's ratio, fatigue, compressive properties, energy absorption, shape memory behaviour, electrical conductivity, as well as piezoresistance. It was found that increasing internal angle enabled to improve the auxetic behaviour of RE honeycombs. In particular, the highest NPR of –1.9 was reached at an internal angle of 75°. The radius in the intercellular connections along with NPR significantly affected the fatigue limit of auxetic structures where a longer fatigue limit was obtained by auxetic structures with higher NPR. Hybrid cellular (HC) cells are made from different unit cells of auxetic structures in order to improve mechanical properties of HC. For instance, auxetic struts constructed from RE honeycomb structures have compressive strength threefold as much as RE honeycomb structures. Another HC built from conventional honeycombs and RE honeycombs called AuxHex may increase their compressive strength, SEA and specific modulus. Deformation temperature T_{def} and peak temperature T_{peak} play an important role in shape memory effect such as strain fixity (SF) and recovery time of the specimens. When T_{def} and T_{peak} were increased, SF values were also enhanced accordingly. Meanwhile, the specimens deformed at lower T_{def} required longer time for recovery. Coating direction influences electrical conductivity of four different lattice structures, namely trust (TR), honeycomb (HN), chiral truss (CT), as well as RE honeycomb (RE) structures made from graphene reinforced with CFs. In the counter direction (CD), all structures except CT exhibited good electrical conductivities since CT has complex structures, enabling to hinder the flow of electrical current. Meanwhile, in the machine direction (MD), the current shows a smooth flow in TR and HN structures, which means the current can flow in both structures without any obstruction.

References

1. Ozturk, M., Baran, T., Tatlier, M.: Experimental and numerical investigation of conventional and stiffened re-entrant cell structures under compression. J. Braz. Mech. Eng. **44**, 593 (2022)
2. Lakes, R.: Design considerations for materials with negative Poisson's ratios. J. Mech. Des. **115**, 696–700 (1993)
3. Li, X., Wang, Q., Yang, Z., Lu, Z.: Novel auxetic structures with enhanced mechanical properties. Extrem. Mech. Lett. **27**, 59–65 (2019)
4. Attard, D., Farrugia, P., Gatt, R., Grima, J.: Starchirals-a novel class of auxetic hierarchal structures. Int. J. Mech. Sci. **179**, 105631 (2020)
5. Xu, J., Wu, Y., Wang, L., Li, J., Yang, Y., Tian, Y., Gong, Z., Zhang, P., Nutt, S., Yin, S.: Compressive properties of hollow lattice truss reinforced honeycombs (Honeytubes) by additive manufacturing: patterning and tube alignment effects. Mater. Des. **156**, 446–457 (2018)
6. Wahl, L., Mass, S., Waldmann, D., Zurbes, A., Freras, P.: Shear stresses in honeycomb sandwich plates: Analytical solution, finite element method and experimental verification. J. Sandwich Struct. Mater. **14**, 449–468 (2012)
7. Zhang, Q., Yang, X., Li, P., Huang, G., Feng, S., Shen, C., Han, B., Zhang, X., Jin, F., Xu, F., Lu, T.: Bioinspired engineering of honeycomb structure—Using nature to inspire human innovation. Prog. Mater. Sci. **74**, 332–400 (2015)
8. Ergene, B., Yalcin, B.: Finite element analyzing of the effect of crack on mechanical behavior of honeycomb and re-entrant structures. J. Polythecnic **23**, 1015–1025 (2020)
9. Nugroho, W.T., Dong, Y., Pramanik, A., Selvan, M.C.P., Zhang, Z., Ramakrishna, S.: Additive manufacturing of re-entrant structures: Well-tailored structures, unique properties, modelling approaches and real applications. Addit. Manuf. **78**, 103829 (2024)
10. Indres, A., Constantinescu, D., Mocian, O.: Bending behavior of 3D printed sandwich beams with different core topologies. Mater. Des. Process. Commun. **3**, e252 (2021)
11. Critchley, R., Hazael, R., Bhatti, K., Wood, D., Peare, A., Johnson, S., Temple, T.: Blast mitigation using polymeric 3D printed auxetic re-entrant honeycomb structures: A preliminary study. Int. J. Prot. Struct. **13**, 469–486 (2021)
12. Zhang, Z., Wen, Q., Li, P., Hu, H.: Application of double arrowhead auxetic honeycomb structure in displacement measurement. Sens. Actuators, A **333**, 113218 (2022)
13. Kolken, E., Zadpoor, A.: Auxetic mechanical metamaterials. RSC Adv. **7**, 5111 (2017)

14. Masters, I., Evans, K.: Models for the elastic deformation of honeycombs. Compos. Struct. **35**, 403–422 (1996)
15. Smith, F., Scarpa, F., Burriesci, G.: Simultaneous optimization of the electromagnetic and mechanical properties of honeycomb materials. In: Paper presented at SPIE's 9th annual international symposium on smart structures and materials, international society for optics and photonics, pp. 582–591 (2002)
16. Qi, D., Yu, H., Hu, W., He, C., Wu, W., Ma, Y.: Bandgap and wave attenuation mechanisms of innovative reentrant and anti-chiral hybrid auxetic metastructure. Extrem. Mech. Lett. **28**, 58–68 (2019)
17. Chen, Y., Fu, M.-H.: Design and modeling of a combined embedded enhanced honeycomb with tunable mechanical properties. Appl. Compos. Mater. **25**, 1041–1055 (2018)
18. Evans, K., Alderson, A.: Auxetic materials: Functional materials and structures from lateral thinking! Adv. Mater. **12**(9), 617–628 (2000)
19. Xue, X., Lin, C., Wu, F., Li, Z., Liao, J.: Lattice structures with negative Poisson's ratio: A review. Mater. Today Commun. **34**, 105132 (2023)
20. Alderson, A., Alderson, K., Attard, D., Evans, K., Gatt, R., Grima, J., Miller, W., Ravirala, N., Smith, C., Zied, K.: Elastic constant of 3-, 4- and 6-connected chiral and antichiral honeycombs subject to uniaxial in-plane loading. Compos. Sci. Technol. **70**, 1042–1048 (2010)
21. Grima, J., Gatt, R., Farrugia, P.: On the properties of auxetic meta-tetrachiral structures. Physica Satus Solidi B **245**, 511–520 (2008)
22. Prall, D., Lakes, R.: Properties of a Chiral Honeycomb with a Poisson's Ratio of −1. Int. J. Mech. Sci. **39**, 305–314 (1997)
23. Lorato, A., Innocenti, P., Scarpa, F., Alderson, A., Alderson, K., Zied, K., Ravirala, N., Miller, W., Smith, C., Evans, K.: The transverse elastic properties of chiral honeycombs. Compos. Sci. Technol. **70**(7), 1057–1063 (2010)
24. Larsen, U.D., Sigmund, O., Bouwstra, S.: Design and fabrication of compliant micromechanisms and structures with negative Poisson's ratio. Microelectromechanical Syst. **6**(2), 99–106 (1997)
25. Qiao, J., Chen, C.: Analyses on the in-plane impact resistance of auxetic double arrowhead honeycombs. J. Appl. Mech. **82**, 1–9 (2015)
26. Lim, T.C.: A 3D auxetic material based on intersecting double arrowheads. Phys. Status Solidi B **253**(7), 1252–1260 (2016)
27. Wang, Y., Wang, L., Ma, Z., Wang, T.: Finite element analysis of a jounce bumper with negative Poisson's ratio structure. Pap. Present. Proc. Inst. Mech. Eng., Part C: J. Mech. Eng. Sci. **231**, 4374–4387 (2017)
28. Ma, L., Chen, Y., Yang, J., Wang, X., Ma, G., Schmidt, R., Schröder, K.: Modal characteristics and damping enhancement of carbon fiber composite auxetic double-arrow corrugated sandwich panels. Compos. Struct. **203**, 539–550 (2018)
29. Zhao, X., Gao, Q., Wang, L., Yu, Q., Ma, Z.: Dynamic crushing of double-arrowed auxetic structure under impact loading. Mater. Des. **160**, 527–537 (2018)
30. Gao, Q., Wang, L., Zhou, Z., Ma, Z.D., Wang, C., Wang, Y.: Theoretical, numerical, and experimental analysis of three-dimensional double-V honeycomb. Mater. Des. **139**, 380–391 (2018)
31. Chow, L., Yick, K., Wong, K., Leung, M., Sun, Y., Kwan, M., Ning, K., Yu, A., Yip, J., Chan, Y., Ng, Z.: 3D printing auxetic architectures for hypertrophic scar. Macromol. Mater. Eng. **307**, 2100866 (2022)
32. Li, T., Chen, Y., Hu, X., Li, Y., Wang, L.: Exploiting negative Poisson's ratio to design 3D-printed composites with enhanced mechanical properties. Mater. Des. **142**, 247–258 (2018)
33. Mousanezhad, D., Babaee, S., Ebrahimi, H., Ghosh, R., Hamouda, A., Bertoldi, K., Vaziri, A.: Hierarchical honeycomb auxetic metamaterials. Sci. Rep. **5**, 18306 (2016)
34. Xu, M., Xu, Z., Zhang, Z., Lei, H., Bai, Y., Fang, D.: Mechanical properties and energy absorption capability of AuxHex structure under in-plane compression: Theoretical and experimental studies. Int. J. Mech. Sci. **159**, 43–57 (2019)

35. Lan, X., Wu, G., Huang, G.: In-plane compression response of foam filled re-entrant auxetic structures. Appl. Compos. Mater. **29**, 2245–2263 (2022)
36. Yang, L., Harrysson, O., West, H., Cormier, D.: Mechanical properties of 3D re-entrant honeycomb auxetic structures realized via additive manufacturing. Int. J. Solids Struct. **69–70**, 475–490 (2015)
37. Leung, M.H., Yick, K., Sun, Y., Chow, L., Ng, S.: 3D printed auxetic heel pads for patients with diabetic mellitus. Comput. Biol. Med. **146**, 105582 (2022)
38. Ghannadpour, S., Mahmoudi, M., Nedjad, K.: Structural behavior of 3D-printed sandwich beams with strut-based lattice core: Experimental and numerical study. Compos. Struct. **281**, 115113 (2022)
39. Farrell, D., McGinn, C., Bennett, G.: Extension twist deformation response of an auxetic cylindrical structure inspired by deformed cell ligaments. Compos. Struct. **238**, 111901 (2020)
40. Xi, H., Xu, J., Cen, S., Huang, S.: Energy absorption characteristics of a novel asymmetric and rotatable re-entrant honeycomb structure. Acta Mech. Solida Sin. **34**, 550–560 (2021)
41. Zhou, Y., Li, Y., Jiang, D., Chen, Y., Xie, Y.M., Jia, L.: In-plane impact behavior of 3D-printed auxetic stainless honeycombs. Eng. Struct. **266**, 114656 (2022)
42. Lavery, N., Cherry, J., Mehmood, S., Davies, H., Girling, B., Sackett, E., Brown, S., Sienz, J.: Effects of hot isostatic pressing on the elastic modulus and tensile properties of 316L parts made by powder bed laser fusion. Mater. Sci. Eng. A **693**, 186–213 (2017)
43. Alomarah, A., Masood, S., Sbarski, I., Faisal, B., Gao, Z., Ruan, D.: Compressive properties of 3D printed auxetic structures: experimental and numerical studies. Virtual Phys. Prototyp. **15**, 1–21 (2019)
44. Hrabe, N., Kircher, R., Quinn, T.: Effects of processing on microstructure and mechanical properties of Ti–6Al–4V fabricated using electron beam melting (EBM): orientation and location. In: Paper presented at proceedings of 23rd solid freeform fabrication symposium, Austin, TX, USA (2012)
45. Hutchings, I., Martin, G.: Inkjet technology for digital fabrication, 1st edn. John Wiley & Sons Ltd., Chichester (2012)
46. Scarpa, F., Tomlin, P.: On the transverse shear modulus of negative Poisson's ratio honeycomb structures. Fatigue Fract. Eng. Mater. Struct. **23**, 717–720 (2000)
47. Whitty, J., Nazaré, F., Alderson, A.: Modelling the effects of density variations on the in-plane Poisson's ratios and young's moduli of periodic conventional and re-entrant honeycombs—Part 1: Rib thickness variations. Cell. Polym. **21**, 69–98 (2002)
48. Alderson, A., Rasburn, J., Beg, S., Mullarkey, P., Perrie, W., Evans, K.: An auxetic filter: A tuneable filter displaying enhanced size selectivity or defouling properties. Ind. Eng. Chem. Res. **39**, 654–665 (2000)
49. Lu, X., Li, X., Yang, Z., Xie, F.: Novel structure with negative Poisson's ratio and enhanced Young's modulus. Composite Structure **138**, 243–252 (2016)
50. Yang, D., Lee, S., Huang, F.: Geometric effects on micropolar elastic honeycomb structure with negative Poisson's ratio using the finite element method. Finite Elem. Anal. Des. **39**, 187–205 (2003)
51. Lee, J., Choi, J., Choi, K.: Application of homogenization FEM analysis to regular and re-entrant honeycomb structures. J. Mater. Sci. **31**, 4105–4110 (1996)
52. Chan, K., Xie, L.: Dependency of densification properties on cell topology of metal foams. Scripta Mater. **48**, 1147–1152 (2003)
53. Ghavidelnia, N., Bodaghi, M., Hedayati, R.: Idealized 3D auxetic mechanical metamaterial: An analytical, numerical, and experimental study. Materials **14**, 993 (2021)
54. Beck, A., Silva, C.: Timoshenko versus Euler beam theory: Pitfalls of a deterministic approach. Struct. Saf. **33**, 19–25 (2011)
55. Senobar, M., Etemadi, E., Nazargah, M.: An analytical investigation of elastic–plastic behaviors of 3D warp and woof auxetic structures. Int. J. Mech. Mater. Des. **17**, 545–561 (2021)
56. Andrews, E., Gioux, G., Onck, P., Gibson, L.: Size effects in ductile cellular solids. Part II: experimental results. Int. J. Mech. Sci. **43**(3), 701–713 (2001)

57. Wang, B., Zhou, C.: The imperfection-sensitivity of origami crash boxes. Int. J. Mech. Sci. **121**, 58–66 (2017)
58. Zhang, Y., Xiang, X., Jing, W., Tengteng, C., Wang, C.: Crushing analysis for novel bio-inspired hierarchical circular structures subjected to axial load. Int. J. Mech. Sci. **140**, 407–431 (2018)
59. Hou, W., He, P., Yang, Y., Sang, L.: Crashworthiness optimization of crash box with 3D-printed lattice structure. Int. J. Mech. Sci. **247**, 108198 (2023)
60. Buck, V., Lopez, C., Nimmegeers, P., Hashem, I., Impe, J.: Multi-objective optimisation of chemical process via improved genetic algorithms: A novel trade-off and termination criteria. Comput. Aided Chem. Eng. **46**, 613–618 (2019)
61. Ma, F., Liang, H., Pu, Y., Wang, Q., Zhao, Y.: Multi-objective optimization of crash box filled with three-dimensional cellular structure under multi-angle impact loading. Proc. Inst. Mech. Eng., Part D: J. Automob. Eng. **235**, 2397–2412 (2021)
62. Choudhry, N., Panda, B., Kumar, S.: In-plane energy absorption characteristics of a modified re-entrant auxetic structure fabricated via 3D printing. Compos. B Eng. **228**, 109437 (2022)
63. Arjunan, A., Zahid, S., Baroutaji, A., Robinson, J.: 3D printed auxetic nasopharyngeal swabs for COVID-19 sample collection. J. Mech. Behav. Biomed. Mater. **114**, 104175 (2021)
64. Etemadi, E., Zhang, M., Li, K., Bashtani, M., Ho, M.M.P., Tahir, D., Hu, H.: Composite Structures. **319**, 117206
65. Singh, D., Tobin, D., Dowling, L., Trimble, D.: Optimization of cobalt chrome 3D re-entrant auxetics fabricated using selective laser melting. Eng. Struct. **278**, 115542 (2023)
66. Carakapurwa, F., Santosa, S.: Design optimization of auxetic structure for crashworthy pouch battery protection using machine learning method. Energies **15**, 8404 (2022)
67. Strek, T., Jopek, H., Idczak, E., Wojciechowski, K.: Computational modelling of structures with non-intuitive behaviour. Materials **10**, 1386 (2017)
68. Bendsoe, M., Sigmund, O.: Topology optimization theory, methods, and applications. Springer, Berlin, Heidelberg (2004)
69. Taherkhani, B., Anaraki, A., Kadkhodapour, J., Rezaei, S., & Tu, H.: Large deformation of TPU re-entrant auxetic structures designed by TO approach. J. Elastomers Plast. **53** (2020)
70. Kaminakis, N., Stavroulakis, G.: Topology optimization for compliant mechanisms, using evolutionary-hybrid algorithms and application to the design of auxetic materials. Compos. B Eng. **43**, 2655–2668 (2012)
71. Lukasiak, T.: Two-phase isotropic composites with prescribed bulk and shear moduli. In: Lodygowski, T., Rakowski, J., Litewka, P. (Eds.), Recent advances in computational mechanics, (pp. 213–222). London, UK: Taylor & Francis Group (2014)
72. Andreassen, E., Lazarov, B., Sigmund, O.: Design of manufacturable 3D extremal elastic microstructure. Mech. Mater. **69**, 1–10 (2014)
73. Wang, F.: Systematic design of 3D auxetic lattice materials with programmable Poisson's ratio for finite strains. J. Mech. Phys. Solids **114**, 303–318 (2018)
74. Pasini, C., Inverardi, N., Battini, D., Scalet, G., Marconi, S., Auricchio, F., Pandini, S.: Experimental investigation and modeling of the temperature memory effect in a 4D-printed auxetic structure. Smart Mater. Struct. **31**, 095021 (2022)
75. Choi, H., Shin, E., Lee, S.: Design and evaluation of 3D-printed auxetic structures coated by CWPU/graphene as strain sensor. Sci. Rep. **12**, 7780 (2022)
76. Lvov, V., Senatov, F., Korsunsky, A., Salimon, A.: Design and mechanical properties of 3D-printed auxetic honeycomb structure. Mater. Today Commun. **24**, 101173 (2020)
77. Li, S., Hassanin, H., Attallah, M., Adkins, N., Essa, K.: Poisson's ratio structure using selective laser melting. Acta Mater. **105**, 75–83 (2016)
78. Zhang, W., Tian, R., Zhang, X., Wei, F., Yang, X.: A novel butterfly-shaped auxetic structure with negative Poisson's ratio and enhanced stiffness. J. Mater. Sci. **56**, 14139–14156 (2021)
79. Lan, X., Meng, L., Zhao, J., & Wang, Z.: Mechanical properties and damage characterizations of 3D double- arrowhead auxetic structure with high-relative-density realized via selective laser melting. Eur. J. Mech.—A/Solids. **90**, 104386 (2021)
80. Austermann, J., Redmann, A., Dahmen, V., Quintanilla, A., Mecham, S., Osswald, T.: Fiber-reinforced composite sandwich structures by co-curing with additive manufactured epoxy lattices. J. Compos. Sci. **3**, 53 (2019)

81. Ingrole, A., Hao, A., Liang, R.: Design and modeling of auxetic and hybrid honeycomb structures for in-plane property enhancement. Mater. Des. **117**, 72–83 (2017)
82. Choi, J., Lakes, R.: Nonlinear properties of metallic cellular materials with a negative Poisson's ratio. J. Mater. Sci. **27**, 5375–5381 (1992)
83. Kramberger, J., Nečemer, B., Glodež, S.: Assessing the cracking behavior of auxetic cellular structures by using both a numerical and an experimental approach. Theoret. Appl. Fract. Mech. **101**, 17–24 (2019)
84. Ulbin, M., Borovinšek, M., Vesenjak, M., Glodež, S.: Computational fatigue analysis of auxetic cellular structures made of SLM AlSi10Mg alloy. Metals **10**, 945 (2020)
85. Tomažinčič, D., Vesenjak, M., Klemenc, J.: Prediction of static and low-cycle durability of porous cellular structures with positive and negative Poisson's ratios. Theoret. Appl. Fract. Mech. **106**, 102479 (2020)
86. Nečemer, B., Kramberger, J., Vuherer, T., Glodež, S.: Fatigue crack initiation and propagation in re-entrant auxetic cellular structures. Int. J. Fatigue **126**, 241–247 (2019)
87. Sun, L., Huang, W.: Mechanisms of the multi-shape memory effect and temperature memory effect in shape memory polymers. Soft Matter **6**, 4403–4406 (2010)
88. Miaudet, P., Derré, A., Maugey, M., Zakri, C., Piccione, P., Inoubli, R., Poulin, P.: Shape and temperature memory of nanocomposites with broadened glass transition. Science **318**, 1294–1296 (2007)
89. Xie, T.: Tunable polymer multi-shape memory effect. Nature **464**, 267–270 (2010)
90. Yu, K., Dunn, M., Qi, H.: Digital manufacture of shape changing components. Extrem. Mech. Lett. **4**, 9–17 (2015)
91. Kratz, K., Madbouly, S., Wagermaier, W., Lendlein, A.: Temperature-memory polymer networks with crystallizable controlling units. Adv. Mater. **23**, 4058–4062 (2011)
92. Andrew, J., Alhashmi, H., Schiffer, A., Kumar, S., Deshpande, V.: Energy absorption and self-sensing performance of 3D printed CF/PEEK cellular composites. Mater. Des. **208**, 109863 (2021)
93. Mora, A., Verma, P., Kumar, S.: Electrical conductivity of CNT/polymer composites: 3D printing, measurements and modeling. Compos. B Eng. **183**, 107600 (2020)
94. AlMahri, S., Schneider, J., Schiffer, A., Kumar, S.: Piezoresistive sensing performance of multifunctional MWCNT/HDPE auxetic structures enabled by additive manufacturing. Polym. Test. **114**, 107687 (2022)

Chapter 5
4D Printing of Shape Memory PU Composites

Abstract The basic concept of 4D printing and main difference between 4D printing and general 3D printing are covered in this chapter. This chapter also specifically elaborates the preparation of 4D printed polyurethane (PU) composites. It is then followed by the elucidation related to the properties of resulting 4D printed structures. Finally, shape memory effect (SME), external stimuli and associated mechanisms are exclusively detailed as a unique characteristic.

Keywords 4D printing · PU composites · Shape memory effect · External stimuli · Programmed materials

5.1 Basic Concept of 4D Printing

4D printing is basically 3D printing process that includes smart materials, programming, stimuli and time [1]. Smart materials refer to specific materials with typical SME, such as regaining their initial shape when subjected to an external stimulus (e.g. light, heat, magnetic and electrical fields) [1, 2] after being deformed by external force. Programming is employed in relation to shape profile of 3D printed parts where those parts are initially deformed. Then an external stimulus is imposed to facilitate their shape recovery. On the other hand, the term "time" reflects that 4D printed parts can change over time leading to final shape recovery [2]. This key concept was mentioned by Pei [3] that 3D printing could build a static object, while 4D printing incorporated designed geometries where the deposition of smart materials was precisely controlled by allowing the shape of printed parts to change when subjected to external stimuli. Fundamental manufacturing comparison between 3D printing and 4D printing are depicted in Fig. 5.1.

Shape memory PU, which belongs to a family of shape memory materials, may become a good candidate as feedstock materials utilised for 4D printing, along with widespread applications for actuators, sensors, wearable, scaffolds and biomedical devices [5–8]. However, the characteristics of 3D printed shape memory PU still have not met the demand of end-users yet. Hence, it needs to be combined with

W. T. Nugroho et al., *Thermoplastic Polyurethane (TPU)/Halloysite Nanotube (HNT) Nanocomposites*, Engineering Materials, https://doi.org/10.1007/978-981-95-0541-8_5

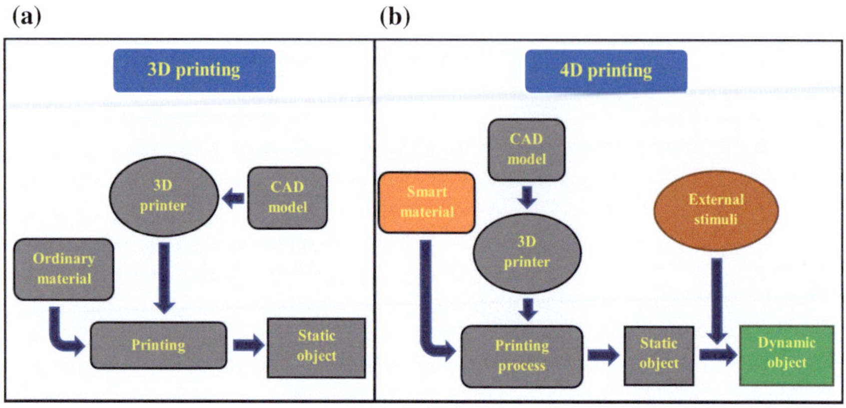

Fig. 5.1 Fundamental manufacturing comparison between **a** 3D printing and **b** 4D printing [4]

other materials like fibres, nanoparticles and polymers to significantly improve its material characteristics. The presence of additives within PU matrices enables the enhancements of SME, self-healing features, electrical properties, sensing capability and biocompatibility of PU composites. Table 5.1 indicates typically used PU, as well as their 4D printed composites.

Fused deposition modelling (FDM) technique is widely utilised to manufacture PU composites owing to its merit in processing various materials, including nanocomposite materials like PU/multi-walled carbon nanotube (MWCNT) nanocomposites, PU/carbon black (CB) nanocomposites and PU/tungsten nanocomposites [9–11]. FDM technology provides a cheaper manufacturing process when compared with PolyJet [12]. Nonetheless, such a technology has common disadvantages in manufacturing 3D printed parts with worse surface quality and lower dimensional accuracy when compared with those produced by PolyJet [11].

5.2 Preparation of 4D Printed Shape Memory PU Composites

The preparation steps of 4D printed PU composites involve purification, dehumidification and mixing processes. Table 5.2 provides a summary of 4D printing for PU composites along with programming processes, as well as associated stimuli. In the meantime, the additives, typically comprising nanoparticles, polymers and microparticles, may be blended with polymer matrices via melt mixing or solution mixing [6, 9–11, 13, 16–18].

In FDM, PU with its additives should be mixed to create filaments by employing a filament extruder. The melting temperature (T_m) of main materials, such as PU with

Table 5.1 Summary of 4D printed PU materials and their composites [4]

PU			Composites			3D printing technique	Ref.
Type	Specification	Supplier	Type	Specification	Supplier		
PU	• MM3520 • Pellets • Glass transition temperature (T_g): 35 °C • Specific gravity: 1.25	• SMP Technologies Inc., Japan	• MWCNTs	• Length: 100–200 μm • Average diameter: 8 nm	• Jeio Co., Incheon, Korea	• FDM	[11]
PU	• DiAPLEX MM 4520 • Pellets • T_g: 45 °C • Specific gravity: 1.25	• SMP Technologies Inc., Japan	• CB	• Ketjenblack, EC600JD	• Akzo Nobel, USA	• FDM	[10]
PU	• MM3520 • Pellets • T_g: 35 °C • Specific gravity: 1.25	• SMP Technologies Inc., Japan	• Tungsten	• Powders • Purity: > 99.9% • Size: 0.6–1 μm	• Sigma Aldrich	• FDM	[9]

(continued)

Table 5.1 (continued)

PU			Composites			3D printing technique	Ref.
Type	Specification	Supplier	Type	Specification	Supplier		
Waterborne PU	• Rust-Oleum triple thick polyurethane clear matt	• Rust-Oleum	• Carbomethyl cellulose (CMC) • Silicon dioxide (SiO2)	• Whatman CMC pre-swollen microgranular cation exchanger CM52 • Size: 10–20 nm • Density: 2.2–2.6 g/mL at 25 °C	• Whatman-Cytiva • Sigma Aldrich	• FDM	[13]
PU	• Elastollan PU, 1185A • T_g: -38 °C • Specific gravity: 1.12 • Melt flow index: 10–20 g/10 min (190 °C, 8.7 kg)	• BASF	• polylactic acid (PLA)	• PLA 8052D • T_g: 60 °C • Specific gravity: 1.24 • Melt flow index: 14 g/10 min (210 °C, 2.16 kg)	• NatureWorks LLC	• FDM	[14]

(continued)

Table 5.1 (continued)

PU			Composites			3D printing technique	Ref.
Type	Specification	Supplier	Type	Specification	Supplier		
PU	• Elastomer PU	• Changsa Yitailong, China	• PLA	• Shape memory material-PLA	• Zhejiang Hailuohao, China	• FDM	[15]
PU	• Filaments • T_g: 55 °C • Filament diameter: 1.75 mm • Melting temperature: 205–215 °C	• SMP Technologies Inc., Japan	• Nylon	• Plain woven fabrics • Thickness: 0.1 mm		• FDM	[16]

Table 5.2 Summary of 4D printing of PU composites [4]

Method	Matrix material	Reinforcement	Material preparation process	Printing parameter	Stimulus	Programming process	Ref.
FDM	• PU with T_g of 35 °C	• MWCNTs	• Solution mixing • Ultrasonication • Stirring • Vaporisation • Shredding • Drying • Melt mixing • Extrusion	• Print speed: 40 mm/s • Nozzle temperature: 200 °C • Bed temperature: 70 °C	• Heat	• Heating the sample above their T_g (40 °C) • Applying external force to deform the sample • Maintaining this deformation at below T_g (20 °C) • Releasing the applied force • Immersing the samples into water at 40, 45 and 50 °C	[11]
FDM	• PU with T_g of 45 °C	• CB	• Solution mixing • Sonication • Evaporation • Cutting • Melt mixing • Extrusion	• Print speed: 80 mm/s • Nozzle temperature: 230 °C • Bed temperature: 60 °C • Layer height: 0.25 mm • Infill: 100%	• Electrical current	• Applying external force and electrical current to deform the samples up to their ultimate tensile strength • Maintaining this deformation • Releasing the applied force and turn off the electrical current after the sample reaches its fracture point	[10]

(continued)

Table 5.2 (continued)

Method	Matrix material	Reinforcement	Material preparation process	Printing parameter	Stimulus	Programming process	Ref.
FDM	• PU with T_g of 35 °C	• Tungsten	• Ball milling • Solution mixing • Stirring • Drying • Melt mixing • Extrusion	• Print speed: 20 mm/s • Nozzle temperature: 230 °C • Bed temperature: 55 °C • Layer height: 0.2 mm • Infill: 100%	• Heat	• Heating the sample above their T_g (60 °C) • Compressing the sample until it is deformed to the half of its thickness • Maintaining this deformation and the sample is cooled at -10 °C • Releasing the applied force • Immersing into water at 40 °C	[9]
FDM	• PU with T_g of 55 °C	• Nylon	• Extrusion	• Print speed: 50 mm/s • Nozzle temperature: 230 °C • Bed temperature: 25 °C • Layer height: 0.2 mm • Infill: 100%	• Heat	• Heating the sample at 70 °C for 5 min • Deforming the object shape • Cooling the sample for 30 min • Reheating the sample at 70 °C for 5 min	[16]

(continued)

Table 5.2 (continued)

Method	Matrix material	Reinforcement	Material preparation process	Printing parameter	Stimulus	Programming process	Ref.
FDM	• PU with melting temperature (T_m) of 170–220 °C	• poly (ε-caprolactone) (PCL) • MWCNTs	• Solution mixing • Stirring • Washing • Drying • Creating composite films • Crushing the composite films • Extrusion	• Print speed: 30 mm/s • Nozzle temperature: 230 °C • Bed temperature: 30 °C • Layer height: 0.4 mm • Infill: 100%	• Light (near infrared)	• Heating the sample above T_m -PCL • Deforming the object shape • Cooling the sample • Applying near infrared (NIR) to reheat the sample	[17]
FDM	• PU with T_g of -20 °C	• Wood flour (WF)	• Mechanical mixing • Melt mixing • Extrusion	• Print speed: 25 mm/s • Nozzle temperature: 230 °C • Bed temperature: 30 °C • Layer height: 0.2 mm • Infill: 100% • Filling angle: 45°	• Heat	• Heating the sample above T_m-PU for 2 h • Deforming the sample • Cooling the sample at -25 °C for 12 h • Reheating the sample above T_m -PU	[18]

(continued)

Table 5.2 (continued)

Method	Matrix material	Reinforcement	Material preparation process	Printing parameter	Stimulus	Programming process	Ref.
FDM	• PU filaments	• PLA filaments with T_g in a range of 60-65 °C	• Cutting the filaments to pellet rods • Aeration with the compressed air • Melt mixing • Extrusion	• Print speed: 10 mm/s • Nozzle temperature: 240 °C • Bed temperature: 70 °C • Layer height: 0.3 mm	• Heat	• Heating the sample at 70 °C • Stretching the sample • Cooling the sample at 21 °C for 5 min • Reheating the sample at 70 °C	[6]
Modified FDM	• Waterborne PU	• Carbomethyl cellulose (CMC) • Silicon dioxide (SiO_2)	• FDM modification (FDM nozzle was replaced by a syringe) • Solution mixing • Stirring • Extrusion	• Pump pressure: 3–50 psi • Layer height: 0.5–0.7 mm • Bed temperature: 50 °C	• Heat	• Heating the printed object at 121 °C onto a hot plate • Deforming the object to a new shape • Removing object from the hot plate • Placing the object onto the hot plate at 121 °C	[13]

its SME, must be taken into account when the extrusion process takes place. Otherwise, it results in undesired filaments with rough surfaces and inconsistent diameters, and (or) even damages those filaments. To obtain desired 3D printed structures, printing parameters, consisting of print speed, layer height, printing temperature, infill density and infill pattern, are highly recommended for parametric optimisation to achieve required filament characteristics such as uniform diameter with a tolerance within $\pm$ 0.1 mm and excellent surface finish [11].

Due to extraordinary SME, PU is typically considered as feedstock materials for 4D printing. Among their several grades, PUs with good SME are comprised of PU elastomer, polyether and polyester grades. Based on the MatWeb database [19], T_m values of elastomer, polyether, polyester and elastomer grades are in range of 100–230 °C, 118–218 °C and 71–221 °C respectively. As such, the printing temperature of 3D printer is commonly set between 200 and 240 °C [6, 11, 16].

T_g is a critical transition temperature observed between glassy and rubbery states. It plays an important role in the programming process of 4D printing because T_g significantly affects shape memory behaviour of printed composites. In relation to this, the deformation program should be performed above T_g. For example, when PU has T_g of 35 °C, printed parts are suggested to be heated above this temperature level (e.g. 40 °C) [11]. For instance, Kabir [16] set the programming temperature about 75 °C when PU possessed a T_g value of 55 °C.

5.3 Properties of 4D Printed Shape Memory PU Composites

Printing parameters like printing temperature, layer thickness, feed rate and filling ratio can affect the properties of 4D printed PU composites. Such printing parameters were implemented by Ly and Kim [11] to evaluate their influence on the recovery behaviour and electrical conductivity of PU/MWCNT composites via FDM. It was mentioned that recovery time of printed composites highly declined when stimulus temperature was set above their T_g. Regarding this, applied stimulus temperature should not exceed 10 °C below T_g to gain effective recovery time [11]. Higher printing temperature, layer thickness and filling ratio and lower feed rate yielded shorter recovery time and lower electrical resistance. It was observed that recovery time was significantly affected by layer thickness and printing temperature [11]. A higher printing temperature eased the bonding formation between layers into a thicker one filled with more MWCNTs, resulting in an increase in electrical conductivity. On the other hand, high filling ratio and low feed rate caused the same effect on the deposition of MWCNTs into the layers. The best result of printed composites was achieved when printing temperature, filling ratio, feed rate and layer thickness were set to 230 °C, 100%, 40 mm/s and 0.2 respectively. This setting resulted in desired composite structures with the resistance of 0.95 kOhm and recovery time up to 42 s [11].

Important characteristics of printed PU composites, such as elongation, toughness and electrical conductivity, may be improved by incorporating CB particles in PU matrices. Rosales et al. [10] selected 1, 3, 5 and 7 wt% CB and then mixed them with PU with the aid of N-dimethylformamide (DMF) as a chemical solvent using the combination of solution mixing and melt mixing. It also took the electric current as an external stimulus. Overall, the electrical conductivity of printed composites was improved when increasing CB content. The best outcome was obtained when CB content reached 5 wt%. Nevertheless, below such content levels (i.e. 1 and 3 wt% CB), resulting printed composites had lower electrical conductivity. It could be associated with the mixing condition of nanocomposites due to single heating chamber used in a single screw extruder. As a result, such an extruder could only work at a relatively low temperature of 160 °C under the pressure of 0.32 MPa. Increasing CB contents up to 5–7 wt%, printed PU/CB composites demonstrated much higher electrical conductivity [10]. However, composite samples reinforced with 5 wt% CB lost their stiffness and also induced material clogging issues during the deposition process. In contrast, printed PU/CB composites reinforced 7 wt% CB particles were too brittle, thus resulting in a high breaking rate in the extrusion process. Therefore, an additional masterbatch of composites with the CB content of 7 wt% was prepared to overcome a detrimental stiffness issue. Afterwards, they were mixed with PU by using a weight ratio of 1:0.33 in the filament extruder to manufacture composite filaments containing 5 wt% CB particles. It was reported that the electrical stimulus could enhance the toughness of PU/CB composites fourfold when compared with controlled samples. Furthermore, the elongation of composite samples was enhanced from 9.45% to 44% at the same level [10]. It exhibited the superplastic characteristics of printed nanocomposites when an electrical stimulus was imposed throughout the programming process.

Tensile properties and SME of sinusoidal PU/nylon composites via FDM were affected by a printing parameter, namely thickness and cycles [16]. It was found that a sinusoidal pattern resembling an oscillating wave yielded remarkable mechanical behaviour. Five different thicknesses, comprising 0.2, 0.4, 0.6, 0.8 and 1.0 mm, were employed in the sinusoidal pattern. All patterns were printed on nylon fabric surfaces using sample codes of 0.2 SM/NF, 0.4 SM/NF, 0.6 SM/NF, 0.8 SM/NF and 1.0 SM/NF depicted in Fig. 5.2.

It was evidently seen in Fig. 5.3 that the thinner samples could recover much faster than thicker counterparts. The samples with a thickness of 0.2 mm (i.e. 0.2 SM/NF), yielded the greatest recovery rate. It was followed by 0.4 SM/NF, 0.6 SM/NF, 0.8

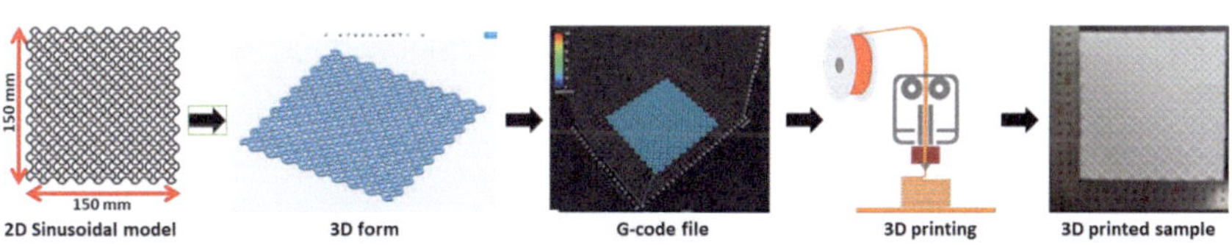

Fig. 5.2 Sinusoidal PU/nylon composites [16]

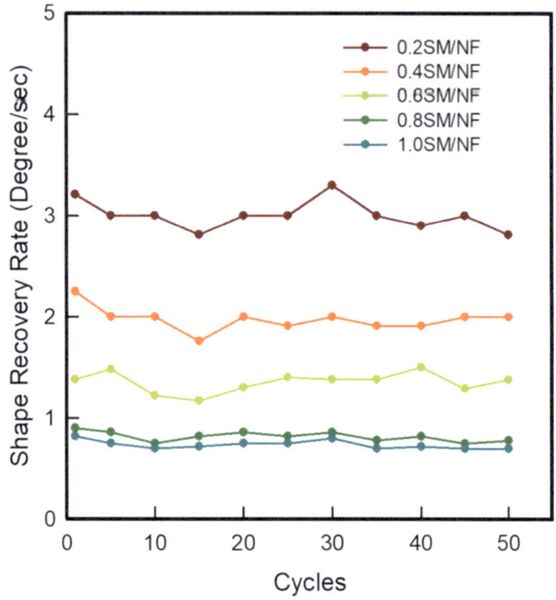

Fig. 5.3 Shape recovery rate of PU/nylon composites with various thicknesses and cycles [16]

SM/NF and 1.0 SM/NF respectively. Furthermore, average shape recovery rates over the period of 50 cycles for 0.2 SM/NF, 0.4 SM/NF, 0.6 SM/NF, 0.8 SM/NF and 1.0 SM/NF were 3.0, 2.0, 1.4, 0.8 and 0.7 /s accordingly. It signified that the thicker samples with more layers required extra time in order to transfer the heat from the exterior to internal layers to reach T_g and shift to a rubbery state, as compared with the thinner samples. Accordingly, the thinner samples were able to set free the stored force and perform recovery more quickly. [20].

5.4 Shape Memory Effect, External Stimuli and Associated Mechanisms

Thermodynamics incompatibility between soft and hard segments in PU generates two different structures with microphase separation. Soft segments are constructed from long-chain polyol, while hard segments are made from diisocyanate and chain extender. In relation to this, hard segments can bind themselves via hydrogen bonding and crystallisation, thus creating PU in a solid form below its T_m. Furthermore, the reversible transformation of soft segments affects SME of material. Five testing methods, consisting of tension, compression, relaxation, flexural and creep tests, can be employed to study SME [21, 22].

Generally speaking, SME, stimuli and associated mechanisms are involved in the study with respect to shape memory materials like PU. Heat was used as a stimulus to activate SME of PU, as mentioned by Kashyap et al. [9]. SME of PU/tungsten composites in size of 10 × 10 × 10 mm was evaluated through programming along with recovery steps in this study. Firstly, the samples were heated up to 60 °C in an oven by referring to T_g of PU (i.e. 35 °C). Subsequently, the samples were compressed to a thickness of 5 mm in one of their dimensions. They were then cooled down to –10 °C in a freezer in order to fix their temporary form. It was followed by the recovery step by immersing the samples in a water bath at 40 °C until they regain their initial form. To assess the shape recovery of such samples, their dimensions in the compression direction were measured. Figure 5.4 shows the recovery process of PU/tungsten composites.

Near-infrared (NIR) light was adopted by Bi et al. [17] to activate SME of dopamine-modified PU/PCL/MWCNT composites via FDM. The presence of MWCNTs was able to absorb NIR light and further convert it into heat to increase the temperature of composite samples, as well as control localised shape change

Fig. 5.4 Shape recovery process of PU/tungsten composites [9]

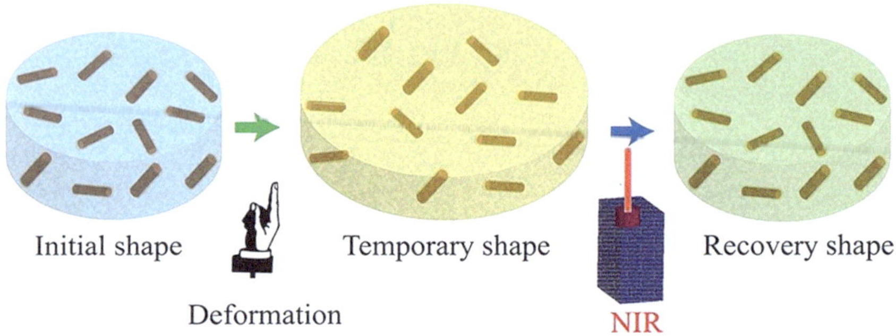

Fig. 5.5 NIR-induced SME process in dopamine-modified PU/PCL/MWCNT composites [17]

[23, 24]. Nonetheless, the interactions between polymer matrices and MWCNTs were quite weak. In contrast, van der Waals forces between those nanoparticles became strong, leading to common particle aggregation. Nanoparticle surfaces should be modified to improve the interfacial interaction between polymer matrices and reinforcements. Accordingly, MWCNTs and NIR were employed as specific materials for photothermal conversion and light source [17]. Figure 5.5 illustrates the programming process to demonstrate SME of nanocomposite structures. During this process, when the temperature level was below T_m, molecular chains of PCL were frozen in a solid-state because of its low molecular movement and crystallisation. Once the temperature was above T_m of PCL, its crystals were slightly soft. Besides, polymeric chains encountered a notable segment movement in order to alter their structures. When PCL was under a temporarily deformed condition, this material remained in a crystalline state with deformed shape. After NIR light was imposed, the temperature of such composites was elevated and further exceeded T_m of PCL with photothermal effect of MWCNTs. This process also softened PCL crystals along with the elastic impact of PU. As such, the form of printed nanocomposites was gradually recovered and then returned to their original shape. In such a composite material system, fixed shape ratio was influenced by high crystallinity and ductility of PCL phase. Conversely, elastic PU phase stored deformation energy and set it free throughout the heating process to soften and restore PCL crystals [25, 26].

5.5 Summary

Notwithstanding that the manufacturing process of 4D printing is similar to 3D printing, 4D printing uses special materials with shape memory characteristics and also involves external stimuli, as well as programming to explore their SME and recovery ability. Purification, dehumidification and mixing processes are particularly required in 4D printing to prepare feedstock materials with minimum impurities and low moisture content. In relation to shape recovery ability of 4D printed parts, thinner samples can recover more quickly than thicker counterparts. PU can be combined with PCL and MWCNTs as feedstock materials for 4D printing by involving NIR light as an external stimulus. Molecular chains of PCL are initially frozen in a solid-state when the temperature level is below T_m owing to its low molecular movement and crystallisation. Nevertheless, PCL crystals become softer after the temperature level is heated above T_m, and polymeric chains undergo segment movement to change their structures. When temporarily deformed imposed by an external force, PCL is in a crystalline state with deformed shape. Once NIR light is used, the temperature level of such composites exceeds T_m of PCL in order to soften PCL crystals and induce the elastic impact of PU towards the recovery process.

References

1. Lee, A., An, J., Chua, C.: Two-way 4D printing: A review on the reversibility of 3D-printed shape memory materials. Engineering **3**, 663–674 (2017)
2. Tibbits, S., Mcknelly, C., Olguin, C., Dikovsky, D., & Hirsch, S.: 4D printing and universal transformation. In: Paper presented at proceedings of the 34th annual conference of the association for computer aided design in architecture, pp 539–548 (2014)
3. Pei, E.: 4D Printing: dawn of an emerging technology cycle. Assem. Autom. **34**, 310–314 (2014)
4. Nugroho, W., Dong, Y., Pramanik, A., Leng, J., Ramakrishna, S.: Smart polyurethane composites for 3D or 4D printing: General-purpose use, sustainability and shape memory effect. Compos. B Eng. **223**, 109104 (2021)
5. Chen, Q., Mangadlao, J., Wallat, J., Leon, A., Pokorski, J., Advincula, R.: 3D printing biocompatible polyurethane/poly(lactic acid)/graphene oxide nanocomposites: Anisotropic properties. ACS App. Mat. Inter. **9**, 4015–4023 (2017)
6. Lathers, S., & Belle, J.: Additive manufactured biomimicking actuator with shape memory polymer composite for prosthetic actuators. 3D Print. Addit. Manuf. **4**, 201–213 (2017)
7. Kim, M., Jung, J., Jung, S., Moon, Y., Kim, D., Kim, J.: Piezoresistive behaviour of additively manufactured multi-walled carbon nanotube/thermoplastic polyurethane nanocomposites. Materials **12**, 1–24 (2019)
8. Joo, H., Cho, S.: Comparative studies on polyurethane composites filled with polyaniline and graphene for DLP-type 3D printing. Polymers **12**, 1–11 (2020)
9. Kashyap, D., Kumar, P., Kanagaraj, S.: 4D printed porous radiopaque shape memory polyurethane for endovascular embolization. Addit. Manuf. **24**, 687–695 (2018)
10. Rosales, C., Duarte, M., Kim, H., Chavez, L., Hodges, D., Mandal, P., Lin, Y., Tseng, T.: 3D printing of shape memory polymer (SMP)/carbon black (CB) nanocomposites with electro-responsive toughness enhancement. Mater. Res. Express **5**, 1–14 (2018)
11. Ly, S., Kim, J.: 4D printing—fused deposition modeling printing with thermal-responsive shape memory polymers. Int. J. Precis. Eng. Manuf. Green Technol. **4**, 267–272 (2017)
12. Gibson, I., Rosen, D., Stucker, B.: Additive manufacturing technologies–3D printing, rapid prototyping, and direct digital manufacturing. Springer Science+Business Media, New York (2015)
13. Su, W., Gao, W., Trinh, K., Kenderes, S., Pulatsu, E., Zhang, C., Whittington, A., Lin, M., Lin, J.: 4D printing of polyurethane paint-based composites. Int. J. Smart Nano Mater. **10**, 237–248 (2019)
14. Jing, X., Mi, H., Peng, X., Turng, L.: The morphology, properties, and shape memory behavior of polylactic acid/thermoplastic polyurethane blends. Polym. Eng. Sci. **55**, 70–80 (2014)

15. Wang, Y., Li, X.: 4D-printed bi-material composite laminate for manufacturing reversible shape-change structures. Compos. B Eng. **2019**, 108918 (2021)
16. Kabir, S., Lee, S.: Study of shape memory and tensile property of 3D printed sinusoidal sample/nylon composite focused on various thicknesses and shape memory cycles. Polymers **12**, 1–17 (2020)
17. Bi, H., Ye, G., Yang, H., Sun, H., Ren, Z., Guo, R., Xu, M., Cai, L., Huang, Z.: Near infrared-induced shape memory polymer composites with dopamine modified multiwall carbon nanotubes via 3D-printing. Eur. Polymer J. **136**, 109920 (2020)
18. Bi, H., Xu, M., Ye, G., Guo, R., Cai, L., Ren, Z.: Mechanical, thermal, and shape memory properties of three-dimensional printing biomass composites. Polymers **10**, 1–14 (2018)
19. Wu, W., Ye, W., Wu, Z., Geng, P., Wang, Y., Zhao, J.: Influence of layer thickness, raster angle, deformation temperature and recovery temperature on the shape-memory effect of 3D-printed polylactic acid samples. Materials **10**, 1–16 (2017)
20. Nejad, H., Baker, R., Mather, P.: Preparation and characterization of triple shape memory composite foams. Soft Matter **10**, 8066–8074 (2014)
21. Feng, X., Zhang, G., Zhuo, S., Jiang, H., Shi, J., Li, F., Li, H.: Dual responsive shape memory polymer/clay nanocomposites. Compos. Sci. Technol. **129**, 53–60 (2016)
22. Chattopadhyay, D., Kothapalli, R.: Structural engineering of polyurethane coatings for high performance applications. Prog. Polym. Sci. **32**, 352–418 (2007)
23. Christ, J., Hohimer, C., Aliheidari, N., Ameli, A., Mo, C., Potschke, P.: 3D printing of highly elastic strain sensors using polyurethane/multiwall carbon nonotube composites. Paper presented at Proceeding SPIE 10168. Sens.S Smart Struct. Technol. Civ., Mech., Aerosp. Syst. **101680E**, 1–7 (2017)
24. Jing, X., Mi, H., Huang, H., Turng, L.: Shape memory thermoplastic polyurethane (TPU)/poly (ε-caprolactone) (PCL) blends as self-knotting sutures. J. Mech. Behav. Biomed. Mater. **64**, 94–103 (2016)
25. Lu, C., Liu, Y., Liu, X., Wang, C., Wang, J., Chu, F.: Sustainable multiple- and multistimulus-shape-memory and self-healing elastomers with semi-interpenetrating network derived from biomass via bulk radical polymerization. ACS Sustain. Chem. Eng. **6**, 6527–6535 (2018)
26. Mosleh, Y., Ebrahimi, N., Mahdavian, A., Ashjari, M.: TPU/PCL/nanomagnetite ternary shape memory composites: studies on their thermal, dynamic-mechanical, rheological and electrical properties. Iran. Polym. J. **23**, 137–145 (2014)

Chapter 6
Methodologies and Analytical Techniques

Abstract The methodologies applied in this study to prepare material samples, consisting of solution casting, extrusion process and 3D printing, are discussed in this chapter. Besides, experimental characterisation, such as the measurements of dimensional accuracy and surface roughness, hardness, abrasion resistance, tensile, shape memory and compression tests, are covered in this chapter. Other characterisations like scanning electron microscopy (SEM), transmission electron microscopy (TEM), Fourier transmission infrared (FTIR) analysis, thermogravimetric analysis (TGA) and differential scanning calorimetry (DSC), are also employed in order to conduct holistic evaluation. To undertake a comprehensive analysis, this study uses statistical design of experiments (DoEs) based on Taguchi method, as well as Pareto ANOVA, along with confirmation tests for parametric optimisation with warranted experimental reproducibility.

Keywords Statistical design of experiments (DoEs) · Mechanical tests · Shape memory tests · Experimental characterisation

6.1 Used Materials

Two different grades of polyurethanes (PUs) were used in this study. The first one is PolyFlexTM PU95 [1] supplied by Polymaker, Shanghai, China in the filament form, which acts as general-purpose PU shown in Fig. 6.1a. The second PU is PU-MM 4520 [2] purchased from SMP Technologies, Tokyo, Japan. This PU obtained in the pellet form is classified as a shape memory material, as depicted in Fig. 6.1b. Halloysite nanotube (HNT) particles selected in this study as the additives in PU/HNT nanocomposites were donated by Imerys Ceramics, Kaeo, New Zealand. Meanwhile, dimethylformamide (DMF) [3] was used as a chemical solvent for dissolving PU pellets to achieve better dispersion of HNT particles, which was supplied by Chem-Supply Pty Ltd., SA, Australia.

W. T. Nugroho et al., *Thermoplastic Polyurethane (TPU)/Halloysite Nanotube (HNT) Nanocomposites*, Engineering Materials, https://doi.org/10.1007/978-981-95-0541-8_6

(a) (b) (c)

Fig. 6.1 **a** PolyFlex™ PU95 filaments, **b** MM 4520 pellets and **c** HNT powders

Table 6.1 Properties of PolyFlex™ PU95 [1]

Physical properties		Mechanical properties	
Density (g/cm^3 at 215 °C)	1.20–1.24	100% modulus (X–Y) (MPa)	9.4 ± 0.3
		Tensile strength (X–Y) (MPa)	29.0 ± 2.8
Melt mass flow rate (g/10 min)	3–6	Elongation at break (X–Y) (%)	330.1 ± 14.9
		Shore D hardness Shore A hardness	46 95

Note (X–Y) denotes testing area subjected to a tensile test. The load direction is parallel to X-axis in sample length, while Y-axis is parallel to sample width.

6.1.1 Thermoplastic Polyurethane (TPU)

TPU belongs to PU family, which is constructed from soft and hard segments in an alternating mode through the reaction between diisocyanates and long-chain diols, and diisocyanates with short-chain diols are often called chain extenders. Various PU types can be made by the modification of structures and molecular weight, as well as mixing ratio between diisocyanates and diols as primary components of PU. The properties of PolyFlex™ PU95 were used as regular PU in this study with material physical and mechanical properties listed in Table 6.1. It should be noted that elongation at break of such PU is approximately 330%, which can be considered as a flexible material.

6.1.2 Shape Memory Polyurethane (SMPU)

Among PU family, shape memory PU (SMPU) is regarded as a unique material with a wider range of T_g, which allows it for shape recovery after being deformed by an external load using an external stimulus such as heat, electrical and magnetic

Table 6.2 Multifunctional properties of PU MM-4520 [2]

Physical properties		Mechanical properties		Thermal properties	
Melt viscosity (Pa·s at 215 °C)	3310	Young's modulus (MPa)	729	Glass transition temperature (°C)	45
		Tensile strength at break (MPa)	41.4		
		Ultimate elongation (%)	600		
		Shore D hardness	72		

fields, as well as light. SMPU is a very promising material candidate for specific applications like actuators, biomedical and wearable devices, scaffolds and sensors [4–7]. PU MM 4520 used in this study possesses shape memory behaviour with multiple properties listed in Table 6.2. It is clearly shown in Tables 6.1 and 6.2 that the maximum elongation of PU MM 4520 is nearly twice PolyFlex™ PU95. Furthermore, SMPU used in this study has greater tensile strength and hardness when compared with PolyFlex™ PU95.

6.1.3 Halloysite Nanotubes (HNTs)

The additives used in this study are halloysite nanotubes (HNTs), which are classified as nanoclays belonging to kaolin family in possession of hollow tubular structures with a high length-diameter ratio. Such materials originate from natural deposits with a chemical formula of $Al_2Si_2O_5(OH)_4{\cdot}nH_2O$. HNT size and its high length-diameter ratio are beneficial for reinforcement effect to polymers such as PU in a nanocomposite system. Moreover, HNTs are abundantly available in nature along with excellent environmental friendliness. As such, HNTs are considered as effective nanofillers in PU nanocomposites. Specific HNT properties of HNTs are listed in Table 6.3.

Table 6.3 HNT properties [8]

Property	Value
Moisture content (%)	3.0
Specific gravity	2.55
pH (aqueous slurry at 20% solids)	3.5–4.5
Surface area [Brunauer–Emmett–Teller (BET)] (m^2/g)	25
Linear shrinkage (dried at 110 °C) (%)	3.8
Modulus of rupture (dried at 110 °C) (MPa)	2.9

6.1.4 *Other Chemicals*

Dimethylformamide (DMF) has a chemical formula of $HCON(CH_3)_2$ with a colourless to pale-yellow characteristic and faint odour. It is deemed an effective chemical solvent used to dissolve PU pellets for better mixing with HNTs to manufacture PU/HNT nanocomposites.

6.2 Sample Preparation

PolyFlexTM PU95 in the filament form was directly utilised by a FDM printer called Axiom 20 (Airwolf 3D, Costa Mesa, USA) to manufacture dog-bone structures based on ASTM D638 standard (type V). On the other hand, MM4520 pellets were mixed with HNTs using DMF as a chemical solvent. Three steps, comprised of the fabrication of PU/HNT nanocomposite films, filaments and samples, were carried out to prepare final PU/HNT nanocomposites according to Fig. 6.2.

PU pellets and HNT particles were firstly dried using a vacuum oven at 80 °C for 4 and 8 h respectively [2, 9] in order to reduce the moisture content. Subsequently, HNT particles were dispersed in DMF with a mix ratio of 1:30 [10] in an ultrasonicating bath at the frequency of 25 kHz with the power intensity of 90% and sonication time of 1 h [11]. On the other hand, PU pellets were dissolved in DMF at a mix ratio of 1:8 [10]. Prior to mixing with PU pellets, DMF was heated using a magnetic stirrer at 300 °C with the rotor speed of 700 rpm. After reaching this temperature level, PU pellets were gradually added into the beaker until those materials were entirely dissolved to avoid material sedimentation.

HNT/DMF mixture was then added into PU/DMF under stirring and heating processes for approximately 30 min. Subsequently, PU/HNT/DMF mixture was cast onto a glass container, and then heated again in an oven at 100 °C for 24 h to remove the solvent. Finally prepared nanocomposite films were stored in a desiccator containing silicon gels prior to further material testing and characterisation. Nanocomposite film fabrication in this study was in good accordance with [12, 13] in the detailed procedure.

The extrusion process at the second step was performed using a Filabot EX6 single screw filament extruder (Filabot company, VT, USA) with the screw diameter of 16 mm and L/D ratio of 24. Four temperature sections of filament extruder comprising feed, back, middle and front sections were set to 45, 165, 165 and 165 °C respectively while screw speed was controlled at 50 rpm in order to create nanocomposite filaments. Before the extrusion process, PU/HNT nanocomposite films were shredded and further chopped into small pieces with the thickness of 0.45 ± 0.1 mm, the width of 3.00 ± 0.1 mm and length of 4.30 ± 0.5 mm accordingly. Subsequently, those pieces were dried in a vacuum oven at the elevated temperature of 80 °C for 4 h to eliminate the moisture. After the extrusion process, resulting nanocomposite filaments were spooled and then stored in airtight containers with silica gels for the

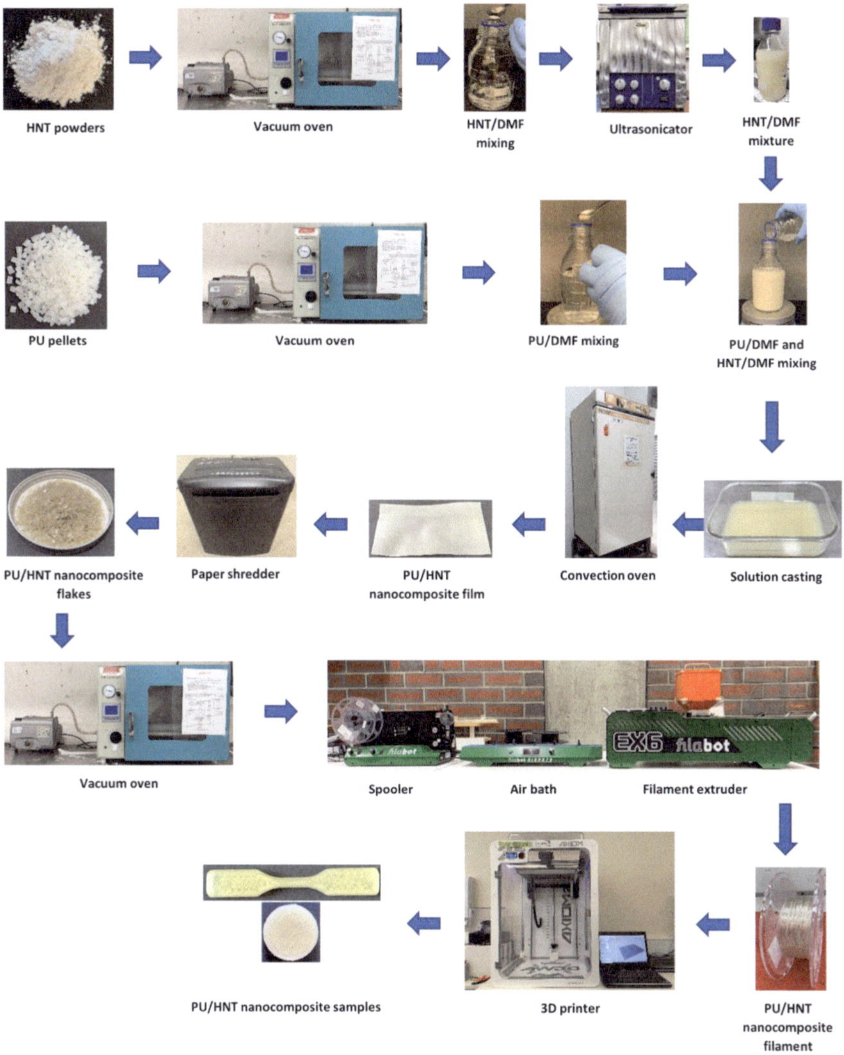

Fig. 6.2 Flow chart of fabricating PU/HNT nanocomposite films, filaments and samples [14]

purpose of material drying. This procedure was also implemented to process virgin PU pellets without shredding and chopping in order to fabricate PU filaments. The extrusion parameters of filament extruder were listed in Table 6.4.

Finally, nanocomposite films were produced by using a FDM based 3D printer Axiom 20 with a built-in software APEX (Airwolf 3D company, USA). These film samples were manufactured based on different HNT contents and settings of selected 3D printing parameters in this study, as shown in Table 6.4. Whereas, other fixed 3D printing parameters are listed in Tables 6.5 and 6.6.

Table 6.4 Summary of different HNT contents and 3D printing parameters in this study

Factor	Level					
HNT content (wt%)	0	2	4	6	8	10
Nozzle temperature (°C)	210		220		230	
Print speed (mm/s)	10		20		30	
Infill density (%)	40		70		100	
Layer height (mm)	0.2		0.3		0.4	

Table 6.5 Fixed 3D printing parameters for original PolyFlex™ PU95 samples

Parameter	Specific parameter	Setting
Quality	Shell thickness (mm)	1.6
	Initial layer thickness (mm)	0.3
	Initial layer line width (%)	100
	Top surface quality	Precise
Fill	Bottom/top thickness (mm)	1.2
	Infill interface density	Dense
	Infill type	Triangle
	Infill overlap (%)	15
Temperature	Bed temperature (°C)	25
Speed	Travel speed (mm/s)	150
	Bottom layer speed (mm/s)	10
	Infill speed (mm/s)	10
Filament	Flow (%)	100
Retraction	Speed (mm/s)	20
	Distance (mm)	20
	Minimum travel (mm)	1.5
	Minimal extrusion before retracting (mm)	0.005

6.3 Experimental Characterisation

Experimental characterisation in this study is comprised of dimensional and surface roughness measurements, tensile, hardness, abrasion resistance, compression and shape memory tests. It is then followed by scanning electron microscopy (SEM), transmission electron microscopy (TEM), Fourier transmission infrared (FTIR), differential scanning calorimetry (DSC), as well as thermogravimetric analysis (TGA) to provide a comprehensive understanding related to the effect of HNT content on mechanical and thermal properties, as well as shape memory behaviour of nanocomposite samples. For test reproducibility, at least three samples were repeated to record required average data with corresponding calculated standard deviations.

Table 6.6 Fixed 3D printing parameters for PU/HNT nanocomposites based on PU-MM4520 and HNTs

Parameter	Specific parameter	Setting
Quality	Shell thickness (mm)	1.6
	Initial layer thickness (mm)	0.5
	Initial layer line width (%)	120
	Top surface quality	Precise
Fill	Bottom/top thickness (mm)	1.2
	Infill interface density	Dense
	Infill type	Triangle
	Infill overlap (%)	15
Temperature	Bed temperature (°C)	55
Speed	Travel speed (mm/s)	150
	Bottom layer speed (mm/s)	15
	Infill speed (mm/s)	30
Filament	Flow (%)	115
Retraction	Speed (mm/s)	30
	Distance (mm)	5
	Minimum travel (mm)	1.5
	Minimal extrusion before retracting (mm)	0.005

6.3.1 Dimensional Accuracy and Surface Finish

Six sample dimensions consisting of thickness, grip section width, full length, parallel section width, circularity and cylindricity were measured via Sheffield Discovery II, a coordinate measuring machine (CMM), along with the standard built-in software PC-DMIS 2013 (Renishaw Electrical Ltd., UK), as illustrated in Fig. 6.3a. This apparatus is equipped with spherical probes in a star configuration. On the other hand, surface roughness of 3D printed samples was examined using Infinite Focus Alicona, an optical 3D measurement instrument (Bruker Alicona, Raaba, Austria) shown in Fig. 6.3b.

The errors of full length (L1-L3), grip section width (W1-W4), as well as thickness (T1-T5) of nanocomposite samples were evaluated by comparing the measurement results of three, four and five different spots with corresponding CAD model. Figure 6.4a depicts the dimensions of CAD model according to ASTM D638 (type V) standard, while Fig. 6.4b shows typical measurement points used for the length and width of the samples.

Meanwhile, the circularity was assessed by measuring eight different spots at the curved middle area [see Fig. 6.4a]. The curved parts of 3D printed sample were split into three regions comprising bottom, middle and top regions to determine its

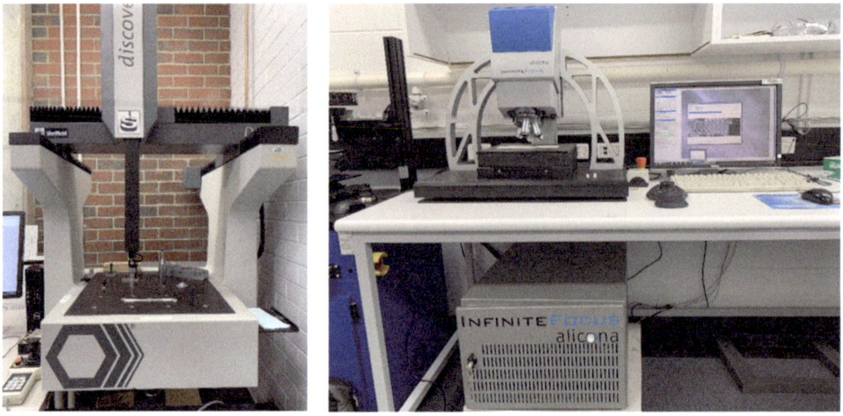

(a) (b)

Fig. 6.3 Physical measurements for **a** dimensions and **b** surface roughness of 3D printed samples

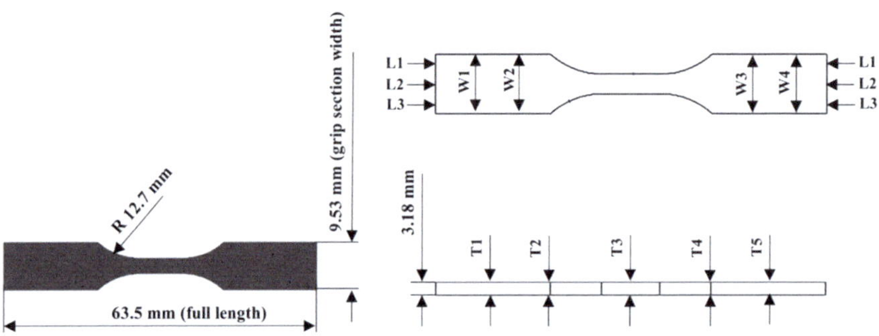

(a) (b)

Fig. 6.4 **a** Dimensions of a dog-bone sample according to ASTM D638 standard (type V) and **b** measurement spots to evaluate the deviations of full length, grip section width and thickness [18]

cylindricity. Four different spots were then measured in each section for accurate data acquisition presented in Fig. 6.4b.

Five modes in PC-DMIS of CMM, namely 'find noms', 'master', 'nominals', 'vector' and 'set nom ax' can be applied to evaluate how the nominals are determined for the specified points to yield the errors of circularity and cylindricity, as illustrated in Fig. 6.5a and b. Regarding this, 'find noms' was chosen for this study due to unknown nominal values. 'Noms' term in 'find noms' reflects nominals, which is a part of analysis technique particularly supplied by PC-DMIS. The deviations of circularity and cylindricity were investigated according to the least-square method [15–17] owing to its mature theory, particularly with simple mathematical formulae and user-friendliness. In addition, it is deemed the most popular technique to evaluate circularity and cylindricity, which is also used as a default mode in PC-DMIS.

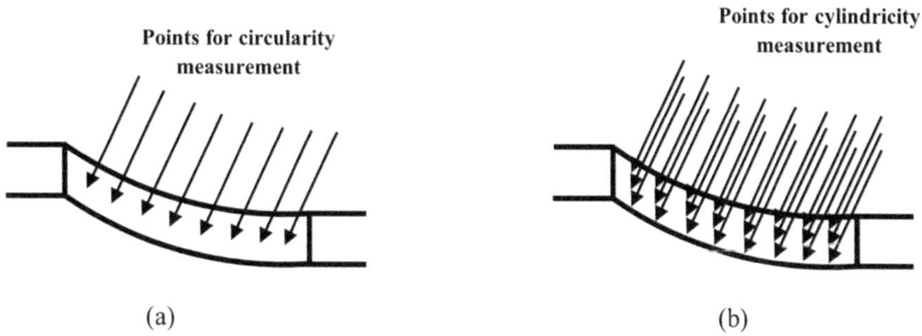

Fig. 6.5 Measurement spots of dog-bone samples for **a** circularity and **b** cylindricity [18]

As mentioned earlier, Infinite Focus Alicona was employed to measure surface roughness of 3D printed samples. This instrument is based on the application of a light source for scanning specific regions to detect surface roughness and surface waviness of nanocomposite samples according to ISO 4287 standard. A selected region was initially scanned, which was then followed by a line scanning step on those desired regions. Subsequently, the data obtained were processed and further analysed via Measure Suit 5.3, a built-in software to determine average surface roughness R_a and waviness W_a of 3D nanocomposite samples. The most critical area of dog-bone samples is the gauge section (marked with red colour) since it is the focussed region for tensile testing, which was divided into three sections, namely top, left and right surfaces in this study according to Fig. 6.6a.

The bottom surface of 3D printed samples was not included in this analysis because it has glossy and smooth surfaces due to the interaction between the first layer under the load (i.e. weight) from the next layers and the build platform originating from tempered glass. As such, it is not facile to distinguish the effect of input (i.e. 3D printing) parameters on output parameters (i.e. surface roughness and surface waviness of nanocomposite samples). The scanning direction should be perpendicular to the raster route of nanocomposite samples [19] according to Fig. 6.6b. Moreover, the measurement path length in scanning process represents the distance between starting and end points.

The measurement was carried out at three different surface areas (i.e. top, left and right surface areas) for each scanning process on the gauge region, consisting of left, middle and right sections shown in Fig. 6.6c. It allowed 3D printed samples to be measured at nine different locations, and the measurement for each sample type was repeated three times. The reference point was taken at the sample surface, which can be indiscriminately located anywhere. For the upper limit, it was adjusted by positioning the light source on the sample surface at the first scanning spot. Afterwards, the device controller was pulled until the surface figure on that spot could not be clearly observed anymore. At this point, the upper limit exceeded ultimate roughness peak. In contrast, the lower limit was set by placing the light source at the end of scanning spot. The device controller was then pressed until the surface figure at that spot became blurry. It is noteworthy that the upper along with the lower

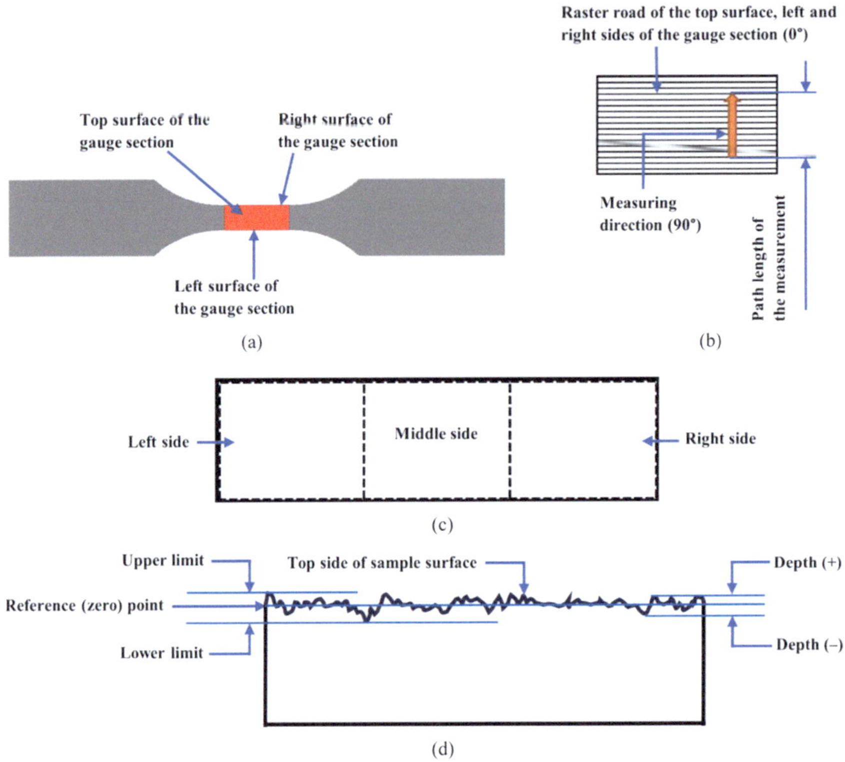

Fig. 6.6 **a** Observed areas for surface roughness and waviness of dog-bone samples, **b** scanning direction, **c** measured areas on the gauge section along with **d** reference point, upper and lower limits. The thickness sample according to CAD model is 3.18 mm, while all dimensions are in mm [18]

limits must be greater than actual roughness values, which were reflected by the depth (+) and (–), as shown in Fig. 6.6d. All measurements, as performed by CMM and optical 3D measurement instrument, were repeated three times to warrant test reproducibility. The resulting average data were calculated for each sample region. These values were employed as the input data for further analysis.

The selected regions for surface roughness analysis with respect to cylindrical samples are comprised of top and side surface sections. Meanwhile, top surface was divided into three areas, namely left, middle and right section areas. On the other hand, side surface consists of four major parts shown in Fig. 6.7. The scanning process on these regions was repeated three times with calculated average data to be used for a subsequent analysis.

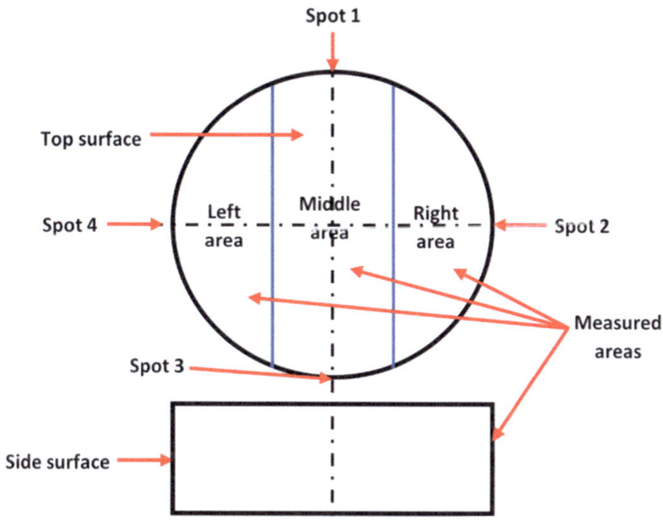

Fig. 6.7 Selected regions for surface roughness analysis of a cylindrical sample

6.3.2 Mechanical Testing

Mechanical properties investigated in this study include tensile properties, namely tensile strength, tensile modulus and elongation at break for 3D printed samples. Other mechanical properties, such as hardness and abrasion resistance, were also identified. All these properties were determined under tensile, hardness and abrasion resistance tests.

6.3.2.1 Tensile Tests

Tensile properties of 3D nanocomposite samples were measured using a Llyod EZ50 universal testing machine (UTM) (Lloyd Instruments Ltd, West Sussex, UK). A typical setup of tensile tests is depicted in Fig. 6.8a with the load cell of 10 kN at the crosshead speed of 5 mm/min. Dog-bone tensile samples were tested at room temperature according to ASTM D638 (type V) shown in Fig. 6.8b.

6.3.2.2 Hardness Tests

HBA 100–0 shore A hardness tester with an analogue mode (Sauter GmbH, Germany) was utilised to determine the hardness of 3D printed virgin PU samples exhibited in Fig. 6.9a. On the other hand, PU/HNT nanocomposites, with recorded hardness data above the capacity of HBA 100–0 shore A hardness tester, were subjected to the hardness measurements using an ICHF-SHRD shore D durometer with a digital

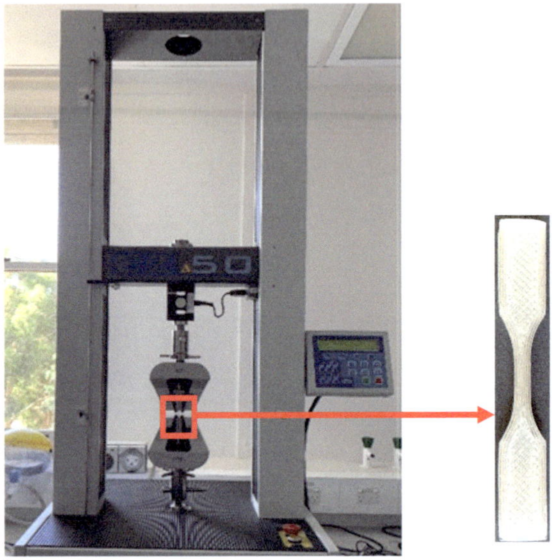

(a) (b)

Fig. 6.8 **a** universal testing machine Lloyds EZ50 and **b** a 3D printed dog-bone sample

mode (Instrument Choice, Australia) shown in Fig. 3.9b. ASTM D2240 standard was followed in the hardness tests. In particular, six spots on the sample surface were used for hardness measurements with a distance as far as 3 mm apart from the sample edge and between individual measurements.

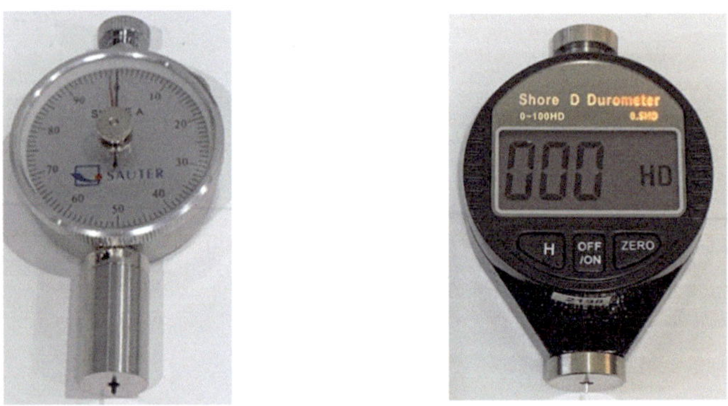

(a) (b)

Fig. 6.9 **a** Sauter analogue durometer HBA 100–0 shore A and **b** digital shore D durometer ICHF-SHRD

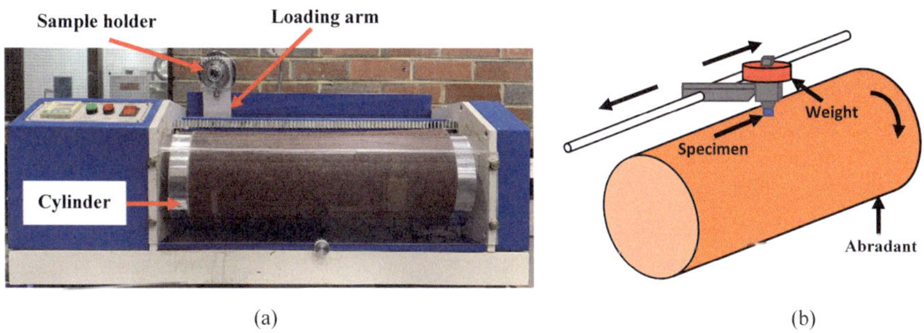

Fig. 6.10 DIN abrasion tester: **a** real setup and **b** corresponding schematic diagram

6.3.2.3 Abrasion Resistance Tests

Abrasion resistance of 3D printed samples was evaluated using their volume loss (mm^3) when those samples underwent abrasion resistant tests using a DZ-323 rotary drum abrader according to ASTM D5963 standard shown in Fig. 6.10a. This instrument consists of a rotating cylinder and a loading arm. At the loading arm, a cylindrical sample was kept by a sample holder in the direction perpendicular to the rotating cylinder surface. The cylindrical sample has the diameter of 16 mm and the thickness of 6 mm. The sample located in the sample holder remained in contact with corundum paper, which is considered as an abrasive medium for two-body abrasion wear, as depicted in Fig. 6.10b.

A normal load was applied via a deadweight system. Volume loss was then determined using the ratio between mass loss and density. The mass loss represents the mass difference of 3D printed samples before and after the abrasion resistance tests. Volume loss was then corrected using S_0/S, in which S_0 and S are referred to as nominal and actual abrasiveness of abrasive paper used in the tests. Additionally, the sample density was determined by using deionised (DI) water according to Archimedes' principle [20].

The rest was conducted on a rotating-test sample by using a standard rubber #1 as a referencing materials based on ASTM D 5963 method B. Mass loss and sample density can be identified by weighing those samples and the referencing material before and after the tests with the aid of a digital scale (accuracy level: $\pm$ 0.1 mg), which is equipped with the density kits (Mettler-Toledo Ltd., Australia). The densities of testing sample and referencing material are defined by Eq. (6.1) where ρ, ρ_0 and ρ_L denote testing sample density, DI density (i.e. $\rho_0 = 1$ g/cm^3) and air density (i.e. $\rho_L = 0.0012$ g/cm^3). Furthermore, A and B represent the weights of a cylindrical sample in air and DI respectively.

$$\rho = \frac{A}{A - B}(\rho_0 - \rho_L) + \rho_L \tag{6.1}$$

Abrasion loss of cylindrical samples is determined using Eq. (6.2)

Table 6.7 Summary of abrasion resistance parameters

Parameter	Value
Sliding distance (m)	40
Sliding speed (m/s)	0.32
Normal load (N)	5
Abrasive medium-corundum (aluminium oxide) (grid)	60

$$A_B = \frac{\Delta m_t \times S_0}{d_t \times S} \tag{6.2}$$

where A_B, Δm_t and d_t are abrasion loss according to ASTM D 5963 (Method B), mass loss, as well as density of a test sample. In addition, S_0 and S represent normal abrasiveness (i.e. $S_0 = 200$ mg) and standard rubber abrasiveness respectively. At the end of the test, wear debris generally in particles on abrasive paper should be removed using a brush. Relevant test parameters were listed in Table 6.7.

6.3.3 *Compression Tests*

Compression tests aim to evaluate the impact of HNT content on compressive properties, including compressive strength, compressive modulus, SEA, as well as PR of 3D printed samples. Re-entrant (RE) samples were used in this study for compression tests, which were made from virgin PU and PU/HNT nanocomposites. RE samples had the length of vertical strut $l = 20$ mm, the length of inclined strut $s = 12$ mm, the height $h = 20$ mm, and the width $w = 20$ mm, the cell thickness $t = 2$ mm, as well as the angle of inclined strut $\theta = 40°$, as depicted in Fig. 6.11a. Similarly, compression tests were also conducted with the load-cell capacity of 10 kN to carry out uniaxial quasi-static tests at room temperature in the in-plane y direction illustrated in Fig. 6.11b. Under these tests, RE samples were set on the fixed platen on the universal testing machine, and three samples were evaluated for test reproducibility at each HNT content.

The crosshead speed of movable platen travelling downwards was set to 0.5 mm/min according to ASTM C365. Test termination took place immediately when a densification phenomenon was observed. Nominal stress σ and normal strain ε are defined according to Eqs. (6.3) and (6.4) as follows

$$\sigma = \frac{F}{l \times w} \tag{6.3}$$

$$\varepsilon = \frac{\delta}{h} \tag{6.4}$$

where F and δ denote the applied load and the displacement change between movable and fixed platens respectively. The built-in software to the universal testing

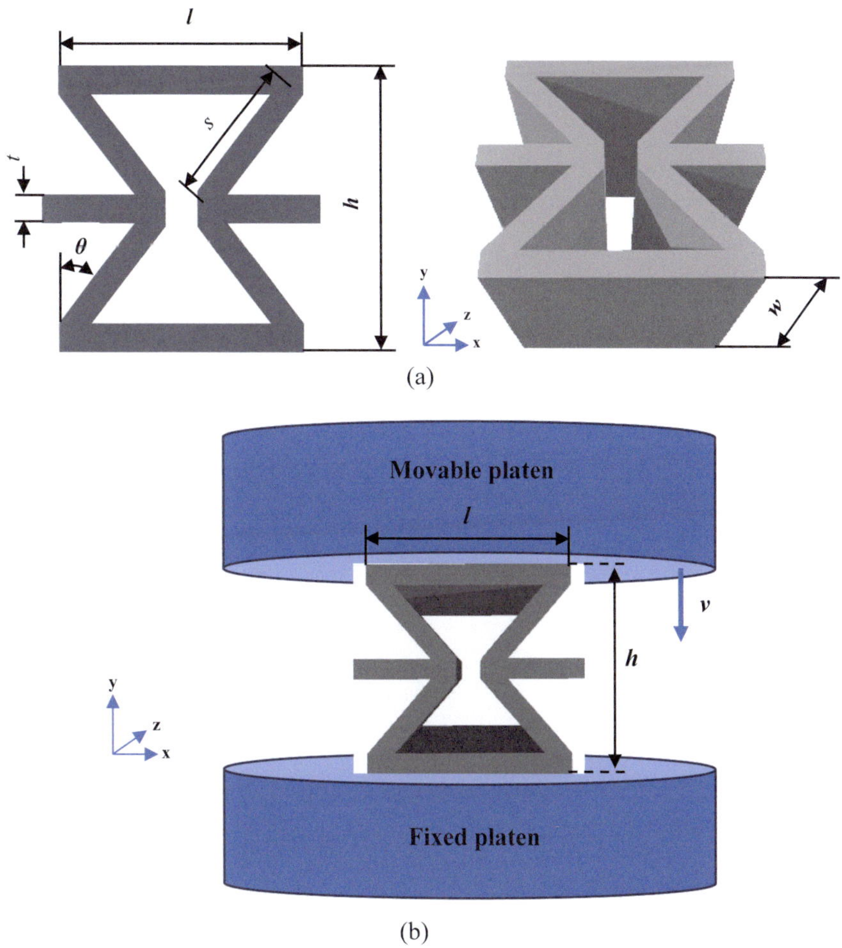

Fig. 6.11 **a** A RE sample and its dimensions (in mm), as well as **b** a schematic diagram of typical compression test for a RE sample

machine was employed to automatically record the displacement, time and force in compression tests.

Meanwhile, compressive data were also used to plot stress–strain curves where the maximum compressive strength of RE samples was obtained at the highest stress level on these curves. Accordingly, onset densification strain is represented by ε_d, which becomes a critical factor for the determination of SEA. As mentioned earlier [21–23], energy absorption efficiency $\eta(\varepsilon)$ should be initially defined using Eq. (6.5) in order to calculate ε_d, as given by

$$\eta(\varepsilon) = \frac{\int_0^{\varepsilon_a} \sigma(\varepsilon) d\varepsilon}{\sigma(\varepsilon_a)} \tag{6.5}$$

in which ε_a is defined as the strain at σ, while η represents the specific region below a stress–strain curve, which is divided by instantaneous stress σ. Furthermore, η is plotted in terms of ε, while ε_d indicates the strain associated with ultimate η, which is written by

$$\frac{d\eta(\varepsilon_a)}{d\varepsilon}|\varepsilon_a = \varepsilon_d = 0 \tag{6.6}$$

SEA represents a cumulative area below the stress–strain curve from 0 to ε_d, which is divided by RE sample density ρ according to Eq. (6.7). The trendline equation should be firstly determined and further integrated to yield the cumulative area under the curve by employing MATLAB R2023a. As aforementioned in previous studies [24, 25], ρ can be determined according to Eq. (6.8) where ρ_s is referred to as bulk material density. As such, bulk material densities of virgin PU and PU/HNT nanocomposites were calculated with the aid of density kits by using Eq. (6.1). It is worth mentioning that θ value of a RE unit cell should be negative according to Eq. (6.8) [24].

$$\text{SEA} = \frac{\int_0^{\varepsilon_d} \sigma(\varepsilon)d\varepsilon}{\rho} \tag{6.7}$$

$$\frac{\rho}{\rho_s} = \frac{\frac{t}{s}\left(\frac{l}{s} + 2\right)}{2\cos\theta\left(\frac{l}{s} + \sin\theta\right)} \tag{6.8}$$

$$\varepsilon = \frac{\Delta h}{h} \tag{6.9}$$

$$\varepsilon_L = \frac{\Delta l}{l} \tag{6.10}$$

$$\text{PR} = -\frac{\varepsilon_L}{\varepsilon} \tag{6.11}$$

The negative ratio of lateral strain to axial strain is defined as Poisson's ratio (PR). The axial strain was determined according to the platen displacement in the y-axis. Conversely, Fujifilm X-T3, a digital camera supplied from Fujifilm Corporation, Tokyo, Japan, was applied to capture RE sample displacement in the x-axis. The resulting data were used to calculate the lateral displacement in the x-axis with the support of DaVinci Resolve 18 and ImageJ 153 software according to Fig. 6.12. The values of axial strain ε and lateral strain ε_L were determined using Eqs. (6.9) and (6.10) where h and Δh denote initial height and the change of height respectively. In addition, l and Δl represent initial length and the change of length. Subsequently, PR is calculated using Eq. (6.11) suggested elsewhere [26].

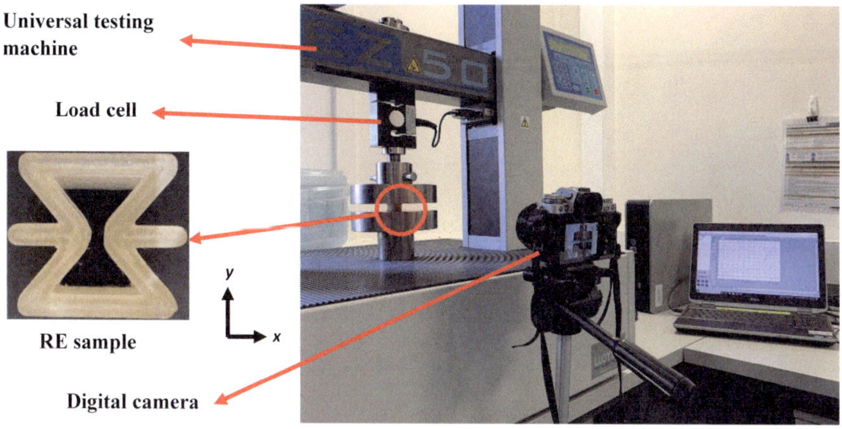

Fig. 6.12 A compression test setup for a unit cell of RE sample

6.3.4 Shape Memory Tests

Shape memory tests are comprised of two steps, namely first (preliminary) and second (main) test steps. In the first step, three-point bending tests were carried out, which were followed by recovery processes to obtain shape ratio and recovery time for deformed samples. This test setup aimed to detect precise temperatures to heat up nanocomposite samples under both compression tests and recovery processes. Such temperature levels were implemented in the second test step. In particular, strip-like samples used in the first test step were prepared according to ASTM D790-17 with their specific dimensions, including the length of 50.8 mm, the width of 12.7 mm, as well as the thickness of 1.5 mm, as depicted in Fig. 6.13. These prepared samples were particularly used for shape memory tests rather than the determination of flexural properties such as flexural strength. A UTM-25 universal testing machine with an integrated oven (IPC Global, Milan, Italy) was employed with the load cell of 25 kN at the crosshead speed of 1 mm/min.

Prepared samples were initially bent at the temperature of 50 °C beyond the glass transition temperature T_g of PU at approximately 41 °C. It is worth mentioning that a twisting phenomenon took place when heated up above 50 °C. A digital thermometer Digitech QM1601 (Digitech, Shandong Province, China) equipped with a K-type thermocouple was used to measure the temperatures of prepared samples in the tests. The recovery process was carried out using a convection oven (LABEC, Laboratory Equipment Pty. Ltd, Marrickville, Australia). Furthermore, shape recovery process was captured via a digital Fujifilm X-T3 camera presented in Fig. 6.13. Three samples were investigated for each material batch at different temperatures of 60, 70 and 80 °C for data reproducibility in order to better understand shape memory behaviour of nanocomposite samples.

Based on ASTM D790-17 standard, it should be noted that the loading must be stopped when midspan deflection D reaches the limit value according to Eq. (6.12).

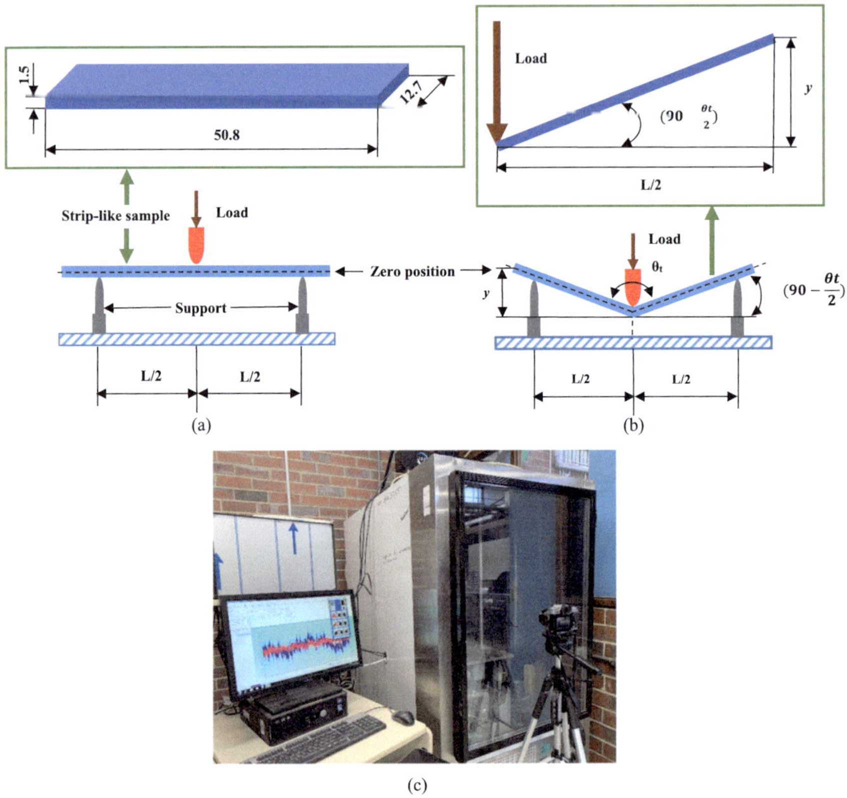

Fig. 6.13 Three-point-bending tests: **a** before loading and **b** after loading, **c** a universal testing machine UTM-25 with an integrated oven

r, L and d hereby represent strain, support span and depth of sample thickness respectively (i.e. $r = 0.05$ mm/mm, $L = 25.4$ mm and $d = 1.5$ mm in this study). After the value input of aforementioned parameters in Eq. (6.12), one can get that the loading appeared to be terminated when $D = 3.58$ mm. In comparison, shape memory ratio R_{rs} used for strip-like samples, as indicated by their angle recovery, enabled to be determined using Eq. (6.13) in which θ_0 and θ_t are recovery angle and programmed bending angle respectively. In particular, θ_0 and θ_t values should be determined at the end of recovery process and loading process respectively.

$$D = \frac{rL^2}{6d} \tag{6.12}$$

$$R_{rs} = \frac{\theta_0 - \theta_t}{\pi - \theta_t} \times 100\% \tag{6.13}$$

The second test step depended on RE samples with similar sample dimensions and compression test procedure at room temperature for characterising shape memory behaviour according to Fig. 6.11 except different sample temperature and testing machines used. Compression tests at the second test step were conducted at 50 °C on an asphalt mixture performance tester (AMPT) with the load cell of 15 kN shown in Fig. 6.14. This equipment has a relatively small load cell when compared with UTM-25 universal testing machine for more accurate data acquisition based on AMPT used for polymer nanocomposites. Additionally, there is a smaller oven for this equipment to reach the desired temperature more rapidly despite the drawback that AMPT is unable to perform three-point-bending tests due to the oven size. This equipment was terminated immediately after $\delta = 8$ mm in which $\delta = 5$ mm was taken into consideration for the onset densification of RE samples. The oven used at the first test step was also implemented to investigate shape memory behaviour of nanocomposite samples at the second test step. The oven heating temperature was set to 80°C because it induced fast recovery and twisting phenomena.

Figure 6.15 demonstrates different heights of RE samples, namely the height before compression h_o, the height after compression h_h, the height after load removal h_i, as well as the height after recovery h_e. Moreover, Eqs. (6.14)–(6.18) can be employed to calculate shape fixity ratio R_f, shape recovery ratio R_r, the amount of deformation after load removal ε_u, maximum strain ε_m, as well as the amount of reversible strain ε_p accordingly.

$$R_f = \frac{\varepsilon_u}{\varepsilon_m} \times 100\% \tag{6.14}$$

$$R_r = \frac{\varepsilon_m - \varepsilon_p}{\varepsilon_m} \times 100\% \tag{6.15}$$

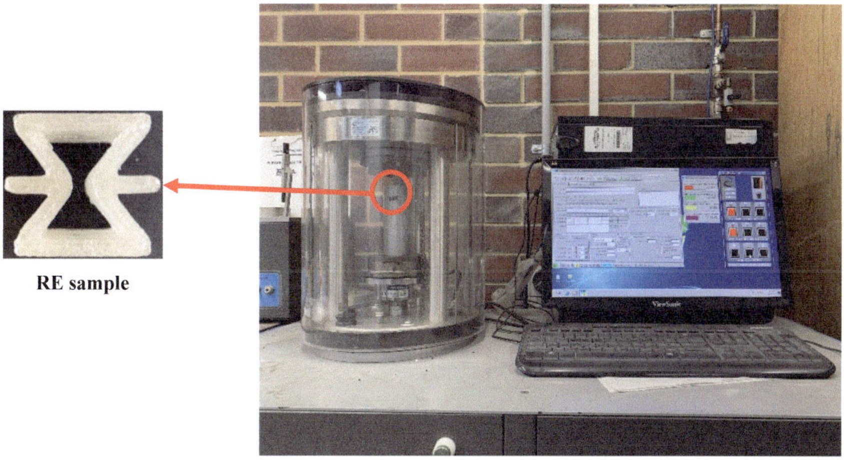

Fig. 6.14 Compression tests for evaluating shape memory behaviour of prepared samples

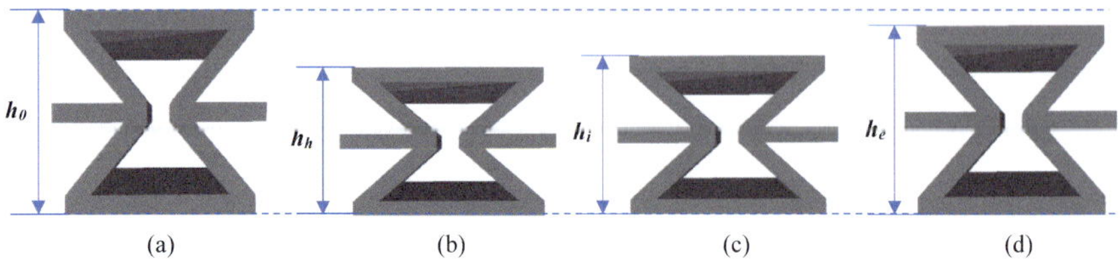

Fig. 6.15 Different heights of RE sample: **a** before compression, **b** after compression, **c** after load removal and **d** after recovery

$$\varepsilon_m = \frac{h_o - h_h}{h_o} \tag{6.16}$$

$$\varepsilon_u = \frac{h_o - h_i}{h_o} \tag{6.17}$$

$$\varepsilon_p = \frac{h_o - h_e}{h_o} \tag{6.18}$$

6.3.5 Characterisation Analysis

A Clara field emission scanning electron microscope (FESEM) (Tescan GmbH, Germany) was utilised to examine surface morphology of nanocomposite samples according to Figure 6.16a at the accelerating voltage of 5 kV and the current of 43 pA. Prior to SEM imaging, material samples were immersed in liquid nitrogen and snapped to prepare clean cross-sectional surfaces in SEM samples. Subsequently, these SEM samples were mounted onto the sample holder of main load (diameter: 45 mm) using double-sided carbon tape. Sputter coating was conducted with carbon layers (layer thickness: 20 mm) on material samples in order to achieve good conductivity and avoid charging issues for better image clarity.

HNT dispersion in PU/HNT nanocomposites was evaluated via a transmission electron microscope (TEM) FEI Talos FS200X G2 with the field emission gun (TEG), as depicted in Fig. 6.16b. Prior to TEM analysis, ultrathin TEM samples (thickness: 100 nm) were prepared using an ultramicrotome Leica EM UC6 with a glass knife and then collected on 300-mesh copper grids.

A thermogravimetric analyser (TGA) SDT 2960 (TA Instruments, USA), shown in Fig. 6.16c, was employed to examine thermal stability of both PU and PU/HNT nanocomposites according to sample weight-loss method. A cooling system based on cryofill liquid nitrogen was operated to assess the weight loss of material samples. About 20 mg TGA samples were heated from 25 to 900 °C at the heating rate of 10 °C/min [12]. On the other hand, a differential scanning calorimeter Discovery DSC 25 (Texas Instruments, USA), as illustrated in Fig. 6.16d, was in operation to

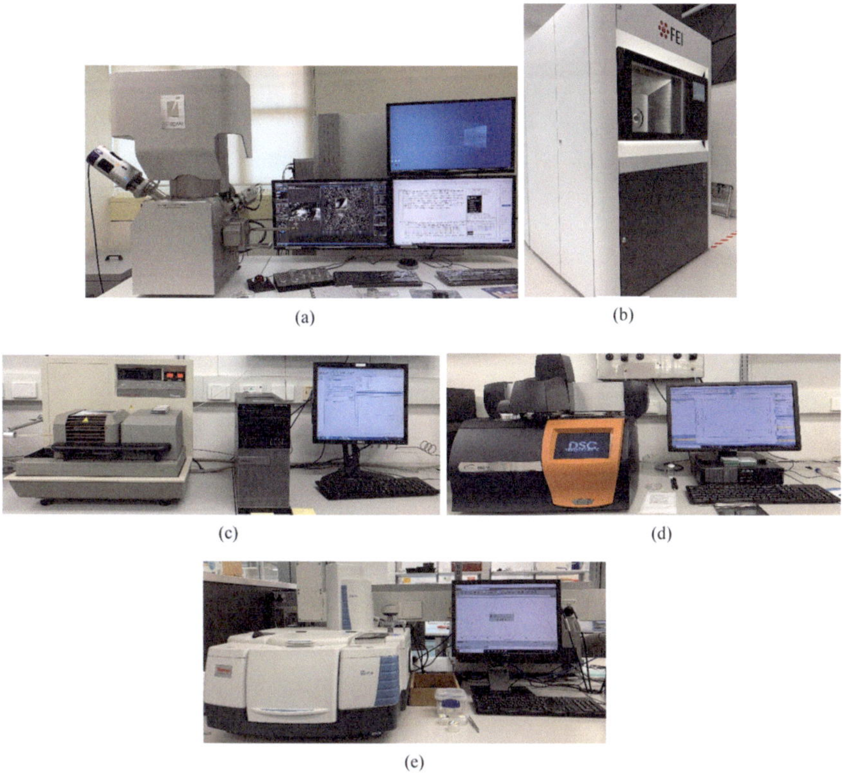

Fig. 6.16 Material characterisation instruments: **a** FESEM-Clara, **b** TEM-FEI Talos FS200X G2, **c** TGA-SDT 2960, **d** DSC-Discovery DSC 25 and **e** FTIR-Nicolet iS50

assess glass transition temperature (T_g), crystallisation temperature (T_c) and melting temperature (T_m) of prepared PU and PU/HNT nanocomposites. About 4 mg DSC samples were sealed in aluminium pans, and further heated from −50 °C to 240 °C at the heating rate of 20 °C/min. An isothermal condition was controlled at 240 °C for 5 min to eliminate thermal history. Afterwards, DSC samples were cooled from 240 °C to –50 °C at the cooling rate of 20 °C/min. Such heating–cooling processes were repeated for the second time. T_g and T_m were determined from the first heating, and T_c was detected from the second heating.

As shown in Fig. 6.16e. Fourier transmission infrared (FTIR) Nicolet iS50 spectrometer (Thermo Fisher, Waltham, USA) was used to evaluate chemical bonds and composition associated with PU and PU/HNT nanocomposite material samples in this study. FTIR analysis was performed at room temperature with an attenuated total reflectance (ATR) technique [27]. Sample spectra were observed in a wavenumber range of 4000 to 400 cm^{-1} with the resolution of 4 cm^{-1}.

6.4 Statistical Design of Experiments (DoEs)

This study used statistical DoEs that involved Taguchi design of experiments and associated orthogonal arrays to assess the impacts of factor-level combinations from input parameters on DoE responses as output parameters. It aimed to achieve an optimum factor-level combination in order to produce material samples with desired material properties. Besides, Pareto analysis of variance (ANOVA) was also implemented to mathematically investigate individual contributions of input parameters.

6.4.1 *Taguchi Method*

Taguchi method is defined as one of most applicable statistical methods to improve manufactured product quality according to Genichi Taguchi [28]. Taguchi method enables to reduce the variation before design optimisation to achieve mean target values with respect to DoE responses due to the ability to optimise design parameters. This method has a wide range of applications in marketing, engineering and biotechnology [29–31].

Taguchi method relies on the orthogonal matrix (array) in experimental design, hereby enabling to evaluate co-influence of some input parameters expeditiously [31]. In general, there are three criteria incorporated into Taguchi method, namely "larger-the better", "nominal-the better", as well as "smaller-the better" criteria. The first two criteria can be used for tensile properties and hardness of material samples, as well as mating parts in the assembly processes. Whereas, "smaller-the better" criterion applies primarily to the assessment of abrasion loss, dimensional accuracy and surface roughness of material samples. More importantly, the optimum level of a factor should be selected based upon mathematically the highest level of sum of signal-to-noise (*S/N*) ratio in experimental design.

6.4.2 *DoE Layout*

Orthogonal arrays (OAs) used in Taguchi method are generally implemented to evaluate all design factors. In this study, a set of fractional factorial designs via Taguchi OAs are required in this study for the assessment of surface quality, dimensional accuracy, tensile strength, tensile modulus, hardness, elongation at break, as well as abrasion resistance of nanocomposite samples. It is worth noting that such a method allows for the optimisation of processing conditions with minimal sensitivities to many variation causes. Additionally, it is less time-consuming and more cost-effective when considering less experimental work, as opposed to a full factorial array in DoE setup [29]. According to Taguchi OAs, the minimum number of experiments can be calculated according to Eq. (6.19) below, as earlier mentioned elsewhere [32]:

$$\text{Number of experiments} = [(L-1)P] + 1 \tag{6.19}$$

where L and P denote the number of levels and parameters respectively. The first study was restricted to four factors comprising print speed, nozzle temperature, infill density and layer height. Those factors were selected as key parameters in 3D printing process with low, intermediate and high level settings.

S/N ratio aims to increase the measurements and quality along with minimising the variability in design and manufacturing [33]. *S/N* ratio applies in this study for response sensitivity measurements in a controlled manner with respect to uncontrolled noise factors. The "smaller-the-better" criterion was chosen in order to minimise DoE responses like surface roughness, abrasion loss and dimensional errors of nanocomposite samples. Conversely, the "larger-the-better" criterion was employed to maximise DoE responses, namely hardness and tensile properties of the same samples. *S/N* ratios for the "smaller-the-better" and the "larger-the-better" criteria were determined using Eqs. (6.20) and (6.21) [34] mentioned below:

$$S/N_i = -10\text{Log}_{10}\frac{1}{n_i}\left(\sum_{u=1}^{n_i} y_u^2\right) \tag{6.20}$$

$$S/N_i = -10\text{Log}_{10}\frac{1}{n_i}\left(\sum_{u=1}^{n_i} \frac{1}{y_u^2}\right) \tag{6.21}$$

in which i, u, n_i and y_u denote the number of experiments, the trial number, the number of trials for ith experiment and the data observed as the DoE response. A higher *S/N* ratio value yields a better result because it warrants better quality with minimum variance. Accordingly, the greatest sum of *S/N* ratio value is important to obtain the optimum factor-level combination.

6.4.3 Pareto ANOVA

ANOVA in this study was employed to assess the contribution of each 3D printing parameter on DoE responses including surface finish, abrasion resistance, hardness and dimension errors, as recommended elsewhere [35]. It is noteworthy that this method does not require an ANOVA table and F-tests to evaluate parameter design results [33]. As such, it becomes less time-consuming for data analysis. Equations (6.22) and (6.23) are implemented to calculate the sum of square (SS) along with the contribution of each factor (P) [34] as follows:

$$\mathrm{SS}_A = \left[\sum_{i=1}^{k_A}\left(\frac{A_i^2}{n_{A_i}}\right)\right] - \frac{T^2}{N} \tag{6.22}$$

$$P_A = \frac{\mathrm{SS}_A - (\nu_A \mathrm{x} \mathrm{V}_e)}{\mathrm{SS}_T} \times 100\% \tag{6.23}$$

where SS_A, A and n_A represent the sum of squares, the value, as well as the number of observations of factor A. Meanwhile, T and N denote the sums of all observations and the total number of observations. The variance due to error, total sum of square of factor A (nozzle temperature), percentage contribution and degree of freedom are represented by V_e, SS_T, P_A *and* v_A. On the other hand, SS and P values for other factors, namely factors B (infill density), C (print speed) and D (layer height), can be calculated according to Eqs. (6.22) and (6.23) respectively. The degree of freedom (dof) of every factor, as indicated by v_i, can be defined as a total of levels subtracted by 1.

6.4.4 Confirmation Tests

In general, DoE work should be validated using confirmation tests by applying significant factors at the optimum factorial level [28]. Such factors are calculated based on the cumulative contribution percentage over 90% [33]. The associated results need to be compared with $S/N_{predicted}$ values based on Eq. (6.24) given below:

$$S/N_{\text{predicted}} = \overline{S/N} + \sum_{j=1} \left(\overline{S/N_k} - \overline{S/N}\right) \tag{6.24}$$

in which $\overline{S/N}$ and $\overline{S/N_k}$ are an average of all *S/N* ratios and average *S/N* ratio in order to calculate significant factors at optimum levels, as well as the selected factor. The quality of confirmation tests relies on the particular requirement where a preferred combination of DoE factors must be confirmed with the use of a confidence interval (CI), which is commonly higher than 95% to improve product quality [28]. Equations (6.25) and (6.26) are utilised to define *CI* for $S/N_{\text{predicted}}$ and $S/N_{\text{confirmation}}$ as follows:

$$\mathrm{CI}_{\text{predicted}} = \sqrt{F_{\alpha,\nu1,\nu2} \times V_e \times \left[\frac{1}{n_{\text{eff}}}\right]} \tag{6.25}$$

$$\mathrm{CI}_{\text{confirmation}} = \sqrt{F_{\alpha,\nu1,\nu2} \times V_e \times \left[\frac{1}{n_{\text{eff}}} + \frac{1}{r}\right]} \tag{6.26}$$

where $F_{\alpha,v1,v2}$, α, v_1 and v_2 are F-ratio obtained from F-table, the risk (i.e. the confidence level is equal to 1-α), dof for the numerator related to a mean (where $v_1 = 1$ in all cases), as well as dof for the denominator (i.e. dof for the pooled error variance) respectively. Meanwhile, n_{eff}, V_e and r represent the effective number of observations, the pooled error variance and the number of confirmation trials (where $n \neq 0$). In this study, confirmation tests were carried out once along with three repetitions for each factor-level combination.

6.5 Summary

Research methodology, as well as analytical techniques used for preliminary study, main study and also application-oriented study have been elaborated in this chapter. It is noteworthy that initial study only employed virgin PU as raw materials to print nanocomposite samples via FDM according to L_9 Taguchi design. The surface quality and dimensional accuracy of those samples were evaluated on an optical 3D measurement device and a coordinate measuring machine. Experimental data acquired in this study was then analysed by Taguchi DoE method and Pareto ANOVA. Afterwards, the data were validated by confirmation tests to investigate the effects of infill density, nozzle temperature, layer height and print speed on dimensional accuracy and surface quality of nanocomposite samples.

The primary study employed PU/HNT nanocomposite filaments as raw materials to manufacture nanocomposite samples via FDM. Accordingly, HNT content, nozzle temperature, print speed, layer height and infill density were considered as input parameters. In contrast, abrasion resistance, tensile properties and hardness were chosen as output parameters (i.e. DoE responses). This primary study also applied Taguchi method using L_{18} design, Pareto ANOVA, as well as confirmation tests to achieve optimum factor-level combinations. Subsequently, nanocomposite samples according to such factor-level combinations were characterised using tensile tests, compression tests, hardness tests, abrasion tests, as well as FTIR, TGA and DSC analyses to assess mechanical, chemical and thermal properties, as well as shape memory behaviour of nanocomposite samples. Additionally, SEM and TEM analyses were applied to examine surface morphology of nanocomposites and HNT dispersion within PU matrices.

RE samples based on PU/HNT nanocomposites were employed for the application-oriented study. The variation of HNT content in range of 0, 2, 4, 6, 8 and 10 wt% applied as the input parameter to investigate its effect on tensile and compressive properties, as well as shape memory behaviour. Finally, the effect of HNT content on specific energy absorption, Poisson's ratio and crack propagation of nanocomposite samples was assessed via compressive tests and SEM analysis.

References

1. Polymaker.: PolyFlexTM TPU95. Polymaker. https://polymaker.com/product/polyflex-tpu95/. Accessed 5 May 5 2021
2. SMP Technologies.: Shape memory polymer. SMP Technologies. , 2022. http://www2.smptechno.com/en/smp/. (2021). Accessed 5 March
3. Scientific, C.S.A.: Dimethylformamide. CSA scientific. https://shop.chemsupply.com.au/search?ProductSearch=DMF. (2021). Accessed Nov 4 2021
4. Seymour, R., Kauffman, G.: Polyurethanes: A class of modern versatile materials. J. Chem. Educ. **69**, 909 (1992)
5. Lathers, S., Belle, J.: Additive manufactured biomimicking actuator with shape memory polymer composite for prosthetic actuators. 3D Print. Addit. Manuf. **4**, 201–213
6. Kim, M., Jung, J., Jung, S., Moon, Y., Kim, D., Kim, J.: Piezoresistive behaviour of additively manufactured multi-walled carbon nanotube/thermoplastic polyurethane nanocomposites. Materials **12**, 1–24 (2019)
7. Joo, H., Cho, S.: Comparative studies on polyurethane composites filled with polyaniline and graphene for DLP-type 3D printing. Polymers **12**, 67 (2020)
8. Imerys.: Halloysite. Imerys. https://www.imerys.com/minerals/halloysite. (2021). Accessed 10 Nov 2021
9. Bouaziz, R., Prashantha, K., Roger, R.: Thermomechanical modeling of halloysite nanotube-filled shape memory polymer nanocomposites. Mech. Adv. Mater. Struct. **26**(14), 1209–1217 (2019)
10. Rosales, C., Duarte, M., Kim, H., Chavez, L., Hodges, D., Mandal, P., Lin, Y., Tseng, T.: 3D printing of shape memory polymer (SMP)/Carbon black (CB) nanocomposites with electro-responsive toughness enhancement. Mat. Res. Express **5**, 1–14 (2018)
11. Salam, H.: Synthesis and characterisation of bio-based epoxy/clay nanocomposites. PhD thesis, Curtin University. (2017)
12. Adak, B., Butola, B., Joshi, M.: Effect of organoclay-type and clay-polyurethane interaction chemistry for tuning the morphology, gas barrier and mechanical properties of clay/polyurethane nanocomposites. Appl. Clay Sci. **161**, 343–353 (2018)
13. Massaro, M., Piana, S., Colletti, C., Noto, R., Riela, S., Baiamonte, C., Giordano, C., Pizzolanti, G., Cavallaro, G., Milioto, S., Lazzara, G.: Multicavity halloysite–amphiphilic cyclodextrin hybrids for co-delivery of natural drugs into thyroid cancer cells. J. Math. Chem. B **3**, 4074–4081 (2015)
14. Nugroho, W.T., Dong, Y., Pramanik, A., Zhang, Z., Ramakrishna, S.: Co-influence of nanofiller content and 3D printing parameters on mechanical properties of thermoplastic polyurethane (TPU)/halloysite nanotube (HNT) nanocomposites. Nanomaterials **13**, 1975 (2023)
15. Sui, W., Zhang, D.: Four methods for roundness evaluation. Phys. Procedia **24**, 2159–2164 (2012)
16. Chen, Q., Tao, X., Lu, J., Wang, X.: Cylindricity error measuring and evaluating for engine cylinder bore in manufacturing procedure. Adv. Mater. Sci. Eng. **2016**, 4212905 (2016)
17. Liu, D., Zheng, P., Wu, J., Yin, H., Zhang, L.: A new method for cylindricity error evaluation based on increment-simplex algorithm. Sci. Prog. **103**, 1–25 (2020)
18. Nugroho, W.T., Dong, Y., Pramanik, A.: Dimensional accuracy and surface finish of 3D printed polyurethane (PU) dog-bone samples optimally manufactured by fused deposition modelling (FDM). Rapid Prototyp. J. **28**, 1770–1795 (2022)
19. Alsoufi, M.S., El-Sayed, A.: Surface roughness quality and dimensional accuracy—a comprehensive analysis at 100% infill printed parts fabricated by a personal/desktop cost-effective FDM 3D printer. Mater. Sci. Appl. **9**(1), 11–40 (2018)
20. IOWA Pressbooks.: Archimedes' principle. https://pressbooks.uiowa.edu/clonedbook/chapter/archimedes-principle/. (2016). Accessed 8 March 8 2022
21. Dong, Z., Li, Y., Zhao, T., Wu, W., Xiao, D., Liang, J.: Experimental and numerical studies on the compressive mechanical properties of the metallic auxetic reentrant honeycomb. Mat. Des. **182**, 108036 (2019)

22. Shen, J., Lu, G., Ruan, D.: Compressive behaviour of closed-cell aluminium foams at high strain rates. Compos. B Eng. **41**, 678–685 (2010)
23. Onck, P., Andrews, E., Gibson, L.: Size effects in ductile cellular solids. part I: modeling. Int. J. Mech. Sci. **43**, 681–699 (2001)
24. Ingrole, A., Hao, A., Liang, R.: Design and modeling of auxetic and hybrid honeycomb structures for in-plane property enhancement. Mater. Des. **117**, 72–83 (2017)
25. Bitzer, T.: Honeycomb technology: material design, manufacturing applications and testing. Chapman & Hall, London, UK (1997)
26. Chow, L., Yick, K., Wong, K., Leung, M., Sun, Y., Kwan, M., Ning, K., Yu, A., Yip, J., Chan, Y., Ng, Z.: 3D printing auxetic architectures for hypertrophic scar. Macromol. Mater. Eng. **307**, 2100866 (2022)
27. Larkin, P.: Infrared and raman spectroscopy: Principles and spectral interpretation, 2nd edn. Elsevier, Stamford (2011)
28. Belavendram, N.: Quality by design, Taguchi techniques for industrial experimentation. Prentice Hall, London, UK (1995)
29. Chohan, J., Singh, R., Boparai, K., Penna, R., Fraternali, F.: Dimensional accuracy analysis of coupled fused deposition modeling and vapour smoothing operations for biomedical applications. Compos. B Eng. **117**, 138–149 (2017)
30. Hong, W.: Using Taguchi method for effective market segmentation. Expert Syst. Appl. **39**, 5451–5459 (2012)
31. Shunmugesh, K., Panneerselvam, K.: Multi-performance optimization of drilling carbon fiber reinforced polymer using Taguchi: Membership function. Trans. Indian Inst. Met. **71**, 1615–1627 (2018)
32. Krishnaiah, K., Shahabudeen, P.: Applied design of experiments and Taguchi methods. New Delhi, India: PHI Learning Private Limited. (2012)
33. Park, S.: Robust design and analysis for quality engineering. Chapman & Hall, London (1996)
34. Ross, P.: Taguchi techniques for quality engineering, loss function, orthogonal experiments, parameter and tolerance design. McGraw-Hill, New York (1988)
35. Quinn, G., Keough, M.: Experimental design and data analysis for biologists. Cambridge University Press, Cambridge, UK (2002)

Chapter 7
Optimisation of Printing Parameters for Dimensional Accuracy and Surface Finish of 3D Printed PU

Abstract This chapter covers dimensional accuracy in terms of full length, grip section width, thickness, circularity and cylindricity, as well as surface quality with respect to surface roughness and surface waviness for PolyFlex™ PU95 dog-bone samples according to ASTM D638 (Type V) standard. Optimal material fabrication of those material samples was achieved with the aid of an Axiom 20 3D printer using a built-in slicer software APEX 1.7.4 in order to fulfil the conversion from a 3D computer-aided design (CAD) model format based on standard triangle language (STL) to G-code prior to 3D printing processes.

Keywords Processing optimisation · Dimensional accuracy · Surface finish · 3D printing

7.1 Identified Significant Parameters

As mentioned earlier in Sect. 6.1.1, PolyFlex™ PU95 is deemed a flexible polymeric material, which is generally more difficult for material processing when compared with much stiffer 3D printed counterparts such as polylactic acid (PLA) and acrylonitrile butadiene styrene (ABS). The nozzle was unable to continuously eject melted filaments onto the print bed when the printing temperature was less than 210 °C at the print speed faster than 12 mm/s. Furthermore, supplied energy is not sufficient to keep the material flow coming out from a nozzle tip, thus it cannot manufacture a complete structure. In contrast, when the printing temperature was higher than 240 °C, printed structures possessed some large voids on their surfaces owing to common overheating issues. On the other hand, such printed structures obtained from PLA and ABS were made with a print speed over 30 mm/s provided that it was not beyond their printing temperature range.

Based on "trial and error" method, the lowest nozzle temperature to fabricate a complete dog-bone sample reached approximately 210 °C, and the print speed must not exceed 12 mm/s. 3D printing parameters, consisting of nozzle temperature, infill density, print speed and layer height, were set to create 3D printed samples,

W. T. Nugroho et al., *Thermoplastic Polyurethane (TPU)/Halloysite Nanotube (HNT) Nanocomposites*, Engineering Materials, https://doi.org/10.1007/978-981-95-0541-8_7

Table 7.1 **3D** printing parameters

Factor	Level		
	Low	Medium	High
A: Nozzle temperature (°C)	210	220	230
B: Infill density (%)	20	60	100
C: Print speed (mm/s)	4	8	12
D: Layer height (mm)	0.1	0.2	0.3

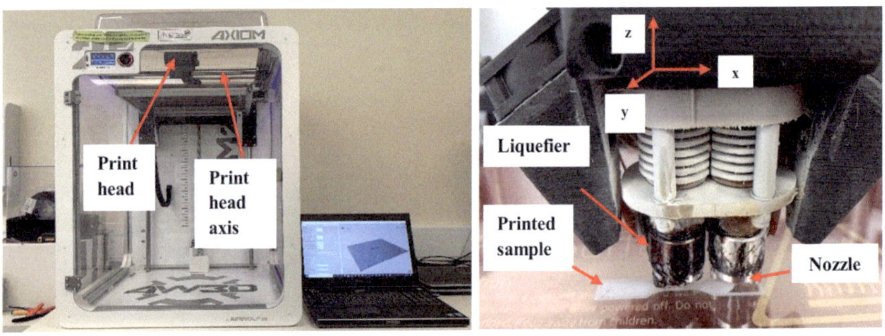

(a) (b)

Fig. 7.1 A **a** 3D printing setup and (b) 3D printing process

as listed in Table 7.1. Nozzle temperature, also called printing temperature, is the temperature of extruding nozzle. Conversely, infill density reflects how full or hollow printed structures are where 0% and 100% mean a shell and a solid respectively. Print speed indicates how quickly the nozzle moves once heated filaments are disposed of on the print bed. Layer height represents the thickness of each layer. The other 3D printing parameters were fixed based on Table 6.5 in Sect. 6.2. The setup of 3D printing parameters and associated process are specified in Fig. 7.1a and b.

3D printed samples were manufactured in the horizontal x axis, which is parallel to the print head axis where z axis refers to the thickness direction shown in Fig. 7.1b. The infill type of nanocomposite samples was in triangular form because it could generate a strong and excellent structure to produce solid layers with top and bottom thicknesses equal to 1.2 mm. Conversely, infill speed was set to 10 mm/s to keep the flow rate of deposited materials onto the print bed, as listed in Table 7.2.

The experimental work performed in this study aimed to investigate major impacts of printing parameters, comprising nozzle temperature, print speed, infill density and layer height, on dimensional deviation and surface quality of nanocomposite structures. Meanwhile, confirmation tests were carried out to assess the validity of resulting data based on design of experiments (DoEs).

According to Eq. (6.19), the lowest number of trials for four input parameters with three levels is 8. Therefore, the closest standard array under this condition should refer to L_9 (3^4) [1], which signifies that 9 trials are required to assess significant

Table 7.2 L_9 design for dog-bone samples according to ASTM D638 (Type V) standard

Exp	Factor				Symbol
	A (nozzle temperature °C)	B (infill density %)	C (print speed mm/s)	D (layer height mm)	
1	210	20	4	0.1	TN1
2	210	60	8	0.2	TN2
3	210	100	12	0.3	TN3
4	220	20	8	0.3	TN4
5	220	60	12	0.1	TN5
6	220	100	4	0.2	TN6
7	230	20	12	0.2	TN7
8	230	60	4	0.3	TN8
9	230	100	8	0.1	TN9

factors used in 3D printing. Table 7.2 displays L_9 OA layout along with the assumption of minimal factorial interaction for simplicity. In addition, the table columns indicate input parameters, and each row reflects a testing condition, originating from a combination of distinct levels of input parameters. Meanwhile, the notation of "TN" in Table 7.2 is referred to as the trial number.

The dimensional accuracy of nanocomposite samples was evaluated using the dimensional deviation, which is then considered as a response to reduce the errors in the determination of signal-to-noise (*S/N*) ratios [2]. A formula to calculate the percentage of deviation (i.e. dimensional error) is expressed using Eq. (7.1) as follows:

$$\text{Percentage of deviation}(\%) = \frac{\text{original CAD dimension} - \text{actual dimension}}{\text{original CAD dimension}} \times 100\% \quad (7.1)$$

It is mentioned in Sect. 6.3.1 that a coordinate measuring machine (CMM) (Sheffield Discovery II), as well as an optical 3D measurement device (Infinite Focus Alicona) were used in this study. The former was implemented to measure sample dimensions, consisting of full length, grip section width, thickness, circularity and cylindricity. The latter referred to surface roughness and waviness of printed samples. Meanwhile, measurement methods and points to calculate those dimensions are detailed in subsection 6.3.1, as also clearly indicated in Figs. 6.4, 6.5, 6.6 and 6. 7.

The experimental data with respect to full length, grip section width, thickness, circularity, cylindricity and surface quality of 3D printed samples are discussed in this chapter. The average values of those measurements were calculated by applying Taguchi DoE method and Pareto ANOVA to assess the effect of 3D printing parameters on dimensional error, surface roughness and waviness of printed samples. Further, confirmation tests were carried out for the validation of DoE results.

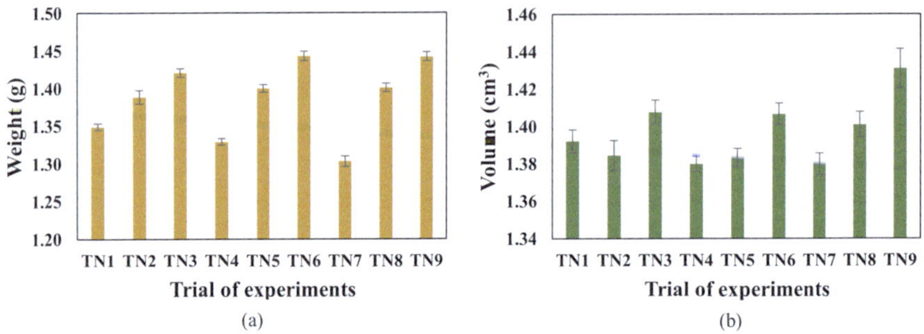

Fig. 7.2 **a** Average weights and **b** average volumes for dog-bone samples

The weight and volume of 3D printed samples were measured using a digital scale and a graduated cylinder. A graduated cylinder is a typical piece of laboratory equipment used to determine the volume of liquid [3]. For each sample type, nine measurements of weight and volume were conducted in this study. Based on the measurement data, average weight and volume for TN3, TN6 and TN9 became greater than those obtained from the rest of samples since these samples were fabricated with an infill density of 100% according to Fig. 7.2. On the other hand, TN2, TN5 and TN8 samples had an infill density of 60%, as opposed to TN1, TN4 and TN7 with the infill density of 20%.

In general, 3D printed samples with the infill density of 100% refer to solid structures without internal voids. TN3, TN6 and TN9 had average weights of 1.42. 1.39 and 1.46 g according to measurement data depicted in Fig. 7.2a. Moreover, average volumes of corresponding samples were 1.41, 1.41 and 1.43 cm^3, respectively, as evidently shown in Fig. 7.2b. As such, sample densities of TN3, TN6 and TN9 were determined to be 1.01, 1.02, and 1 g/cm^3 by dividing their weights by associated volumes. It appeared that the densities of 3D printed samples were lower than those of corresponding solid counterparts when compared with original density of PolyFlex™ PU95 ranging from 1.2 to 1.24 g/cm^3 from the material supplier. This result could be ascribed to the existence of internal pores and voids in 3D printed samples.

Error bars shown in Fig. 7.2a and b refer to standard deviations of average weights and volumes of 3D printed samples. Overall, sample volumes yielded much greater variations than their weights, which may be related to dimensional deviations derived from 3D printing process in longitudinal, transverse and vertical directions, as exemplified by their corresponding errors with respect to full length, grip section width and thickness. Hardness results of PolyFlex™ PU95A filaments were determined using an analogue durometer (Sauter HB model HBA 100–0, Sauter GmbH, Germany) according to ASTM D2240 standard. Average hardness data were obtained by measuring as many as 30 random points around filament surfaces and cross-sectional regions. This resulted in an average shore A hardness of approximately 90, which was slightly lower than that listed in Table 6.1 by the material supplier.

7.2 Thickness, Grip Section Width, and Full Length

Generally speaking, dimensional deviation or dimensional error of 3D printed samples using fused deposition modelling (FDM) against their corresponding CAD models clearly reflects their dimensional accuracy. As illustrated in Fig. 7.3, 3D printed sample thickness induced much greater deviation when compared with those of full length and grip section width. In particular, full length achieved the minimum average deviation of 0.471% with respect to the print orientation. 3D printed samples were fabricated in the flat orientation with the same applied to the direction of print head axis according to Fig. 7.1b to achieve the lowest dimensional errors, which was in good agreement with previous work [4] via FDM for 3D printed PLA dog-bone samples. This work investigated the effect of raster direction, build orientation and layer height on dimensional deviations in relation to thickness, grip section width and weight of the samples. Nonetheless, width deviation appeared to be only slightly lower than that of thickness in the similar build orientation.

In contrast, much greater deviation occurred for sample thickness in good accordance with previous study [5] to investigate the influence of printing temperature, layer height and print speed on the thickness, the length and the width of pyramid structures via FDM. Output parameters refer to the length, the width and the thickness of 3D printed samples in x, y and z directions. Apparently, it was reported that much greater errors appeared for sample thickness in z direction, as opposed to the length and the width. This result may be attributed to the first-layer effect where initial layer was pushed by subsequent layer against the print bed surface to maintain firm structures.

Besides, higher errors of sample thickness could be related to material properties and shrinkage effect. Since PU is generally less stiff than other polymers like PLA, it requires much longer time to be hardened than PLA, as indicated by print

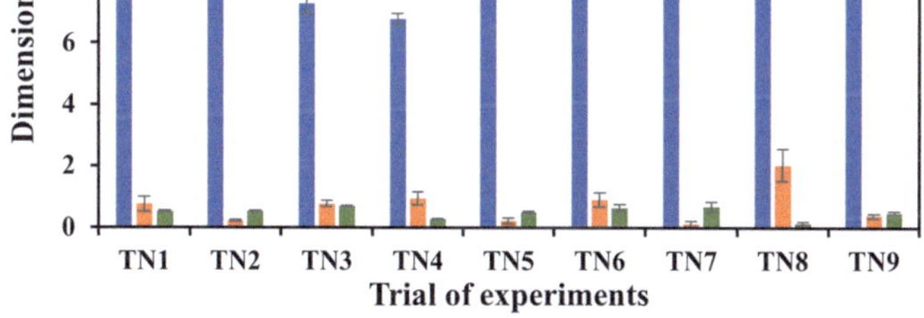

Fig. 7.3 Average dimensional deviations for thickness, grip section width and full length

speed in sample manufacturing. Beyond the print speed of 12 mm/s, the nozzle failed to continuously eject heated PU filaments on the print bed. On the contrary, FDM printing enables to process PLA filaments at the print speed over 90 mm/s, as mentioned elsewhere [6], which signifies that 3D printed PU samples may require much longer manufacturing time than those based on PLA. It is worth mentioning that much longer time for a hardening process yields more shrinkage on 3D printed samples. It could reduce dimensional accuracy of 3D printed samples, especially in the vertical direction represented by sample thickness since such structures were constructed by layer deposition method. The impact of shrinkage in the longitudinal and transverse directions can be reflected by the errors of full length and grip section width of 3D printed samples with average deviations of 0.471% and 0.614% respectively. The length deviation coincided with previous study [5] based on PU, resulting in shore A hardness of approximately 85, which was equivalent to the average deviation of 0.401%. Whereas, this study yielded better accuracy, mainly in the transverse direction with an average width deviation of 0.318%, which was better than 0.837% in width deviation using PLA filaments in previously reported results.

Shrinkage effect in the vertical orientation is generally indicative of thickness error. In this study, average thickness error of samples reached 10.337%. It is worth mentioning that minimum deviation as small as approximately 1.5% was achieved when manufacturing the structures [5]. On the other hand, Hanon et al. [4] obtained a low thickness deviation about 2.432% according to ISO 527–2:2012 standard in 3D printing process. However, these printed samples had much larger sizes than those built in current study. Additionally, the errors of the width and the thickness in the samples could not induce similar effects, which may indicate that the tolerances were still not identical with increasing sample dimensions.

Thickness accuracy of samples might be influenced by the layer number of these samples as well, which yielded higher dimensional errors with more layer number in this study. Layer number was calculated by dividing sample thickness of 3.18 mm according to CAD models by layer thickness. Low, medium and high levels in sample thickness of dog-bone samples, namely 0.1, 0.2 and 0.3 mm, were set according to ASTMD 638 (Type V) standard. As such, corresponding layer numbers became 31.8, 15.9 and 10.6 respectively.

The best accuracy for thickness was obtained when the layer height reached 0.3 mm owing to the fewest layer number, as evidenced by the sum of *S/N* ratios in Fig. 7.4a. The layer height yielded the greatest difference between the maximum and minimum ratios with respect to sample thickness, thereby suggesting its dominant impact on the dimensional deviation of thickness. Additionally, the highest *S/N* ratio was achieved at the layer height of 0.3 mm, resulting in the highest quality with the lowest variance. On the other hand, other factors apparently had minimal effect on sample thickness. Favero et al. [7] evaluated the influence induced by layer height on dimensional deviations of orthodontic models via FDM with similar results. It demonstrated that the greater layer height yielded 3D printed structures with lower deviations. However, it was contradictory with Zhang et al. [8] that decreasing layer thickness gave rise to lower deviations of structures. This finding might be caused by dimensional errors taking place in particular regions of 3D printed structures in each layer, which appeared to be higher with increasing layer number.

The data associated with grip section width were acquired by using a CMM probe to touch sample surfaces on the set spots. As seen from Fig. 7.1b, sample surfaces reflected those shells to wrap all sample sections in the x–y direction, signifying the influence of width errors by shell width. With the shell thickness set to 1.6 mm, shell width could be particularly affected by print speed. Print speed also

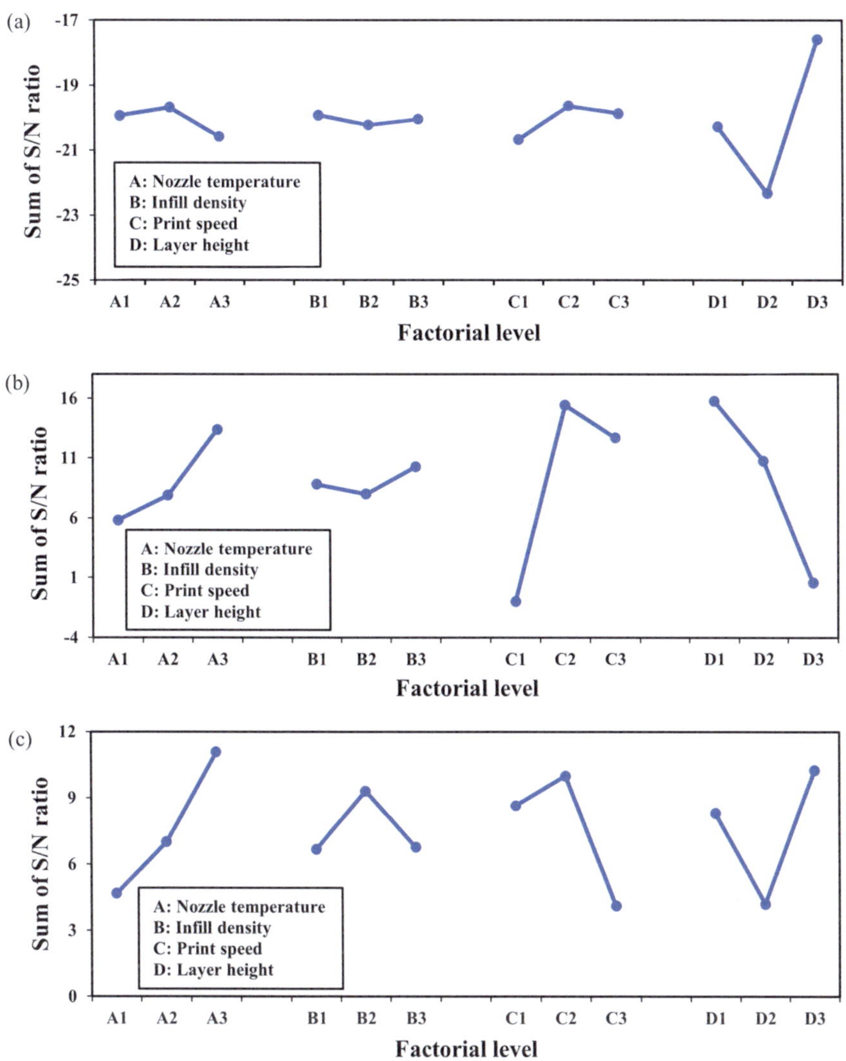

Fig. 7.4 Main effect of sum of *S/N* ratios for 3D printed samples in relation to dimensions according to different factorial levels for **a** thickness, **b** grip section width and **c** full length

played a significant role in grip section width of 3D printed samples according to Fig. 7.4b. In particular, the minimum print speed generated the highest width deviation of 1.236% in relation to manufacturing time to build a layer. In general, longer manufacturing time induces more shrinkage with increasing width deviation. In this study, our experimental data demonstrated that medium print speed gave rise to the minimum deviation of 0.264%, as opposed to the lowest width deviation induced by the maximum speed [9]. Accordingly, optimum print speed might be achieved at the medium level leading to the best width accuracy, which was in good accordance with Žarko et al. [9] using FDM to evaluate the impact of print speed on reproduction quality for embossing tools. It was reported that 3D printed structures made with a higher print speed yielded lower dimensional errors.

Nozzle temperature caused the greatest impact on sample length error when compared with other factors depicted in Fig. 7.4c. A higher nozzle temperature, as indicated by the higher sum of *S/N* ratio, gave rise to the lower deviation. It was detected that average percentages of dimensional error at nozzle temperatures ranging from 210, 220 to 230 °C were 0.590%, 0.473% and 0.349% accordingly. This result could be ascribed to a material-flow issue of extruded filaments at the low temperature of 210 °C to supply insufficient energy without fully melting filament materials in the liquefier and then ejecting them through a nozzle tip. Material flow was increased by elevating printing temperature to obtain lower length errors, which differed from relevant data obtained by Alafaghani et al. [6] to manufacture FDM based PLA structures for the investigation of the influence of building orientation, infill density, printing temperature, print speed, infill pattern, as well as layer height on dimensional error and mechanical properties of 3D printed samples. It was revealed that increasing printing temperature reduced dimensional accuracy. Whereas higher printing temperature caused overflow of heated filaments out of nozzle control, thus lowering dimensional accuracy.

In addition to nozzle temperature, layer height also had remarkable effect on sample length. The minimum full length error could be achieved when the layer height reached a higher level of 0.3 mm, which could be associated with the lowest layer number induced by the largest layer height. Consequently, it caused less shrinkage and the best length accuracy in good accordance with Zhang et al. [8], which was indicative of minimal dimensional deviation of material samples with less layer number. The effect of print speed was also manifested with respect to the accuracy of full length. In particular, a medium level of print speed induced the highest accuracy of full length despite increasing length errors with increasing print speed. A specific temperature gradient appeared at the high print speed of 12 mm/s, which decreased the flow rate of ejected filament materials. As such, it reduced layer length and increased the deviation of sample length. As illustrated in Fig. 7.4, infill density appeared to induce trivial effect on dimensional errors with respect to full length, thickness and grip section width of material samples though different from Gomez-Gras et al. [10] based on polyetherimide (PEI) samples via FDM. It was reported that infill density and thermal effect of 3D printed structures contributed to the shrinkage, thereby resulting in dimensional deviation of 3D printed samples. Consequently, infill density was also considered in a critical role for manufacturing 3D printed structures with desired geometries.

The maximum sum of *S/N* ratio for the factor-level combination with respect to sample thickness was achieved with the nozzle temperature of 220 °C, the infill density of 20%, the print speed of 8 mm/s, as well as the layer height of 0.3 mm according to Fig. 7.4a, which could be represented by $A_2B_1C_2D_3$. On the other hand, the best factor-level combination for grip section width, as shown in Fig. 7.4b, was referred to as $A_3B_3C_2D_1$, namely with the nozzle temperature of 230 °C, infill density of 100%, print speed of 8 mm/s and layer height of 0.1 mm. Finally, the optimum combination for full length was determined to consist of nozzle temperature of 230 °C, infill density of 60%, print speed of 8 mm/s and layer height of 0.3 mm depicted in Fig. 7.4c. Overall contributions of printing parameters to the deviations of full length, thickness and grip section width were presented in Fig. 7.5. Layer height, known as factor D, appeared to be major effect on thickness deviation of 3D printed samples with a contribution percentage beyond 90%, as evidenced by Fig. 7.5a. It confirmed that layer height influenced layer number, further affecting the deviation of sample thickness [7, 8]. With total contribution of approximately 91%, factor D was determined to be the sole significant factor. Furthermore, significant factors for grip section width were comprised of print speed, layer height and nozzle temperature according to Figs. 7.4b and 7.5b.

Print speed affected printing time and shrinkage of 3D printed samples, which induced significant impact on sample width. Additionally, layer height was related to slicing operation in the process of file conversion from STL to G code format. The impact of nozzle temperature with respect to sample width was associated with sample shrinkage effect in a 3D printing process. As far as grip section width was concerned, C_2, D_1 and A_3 could be regarded as the optimum combination for its significant factors shown in Fig. 7.5b.

Full length was influenced by three significant factors comprising nozzle temperature (factor A), layer height (factor D) and print speed (factor C), resulting in an accumulative contribution up to approximately 93% according to Fig. 7.5c. Printing temperature affected material flow of extruded filaments, in which its high level resulted in optimum length accuracy since it yielded better material flow when compared with other levels. Meanwhile, layer height affected required layer number to fabricate a 3D printed sample. In relation to this, the greatest layer height (i.e. 0.3 mm) led to the minimum layer number with the best result. Print speed might affect the flow rate of ejected filaments out of the nozzle tip. The medium level of print speed at 8 mm/s was the optimum owing to resulting maximum length accuracy. The optimum factor-level combination for full-length dimensions could be referred to as A_3, D_3 and C_2.

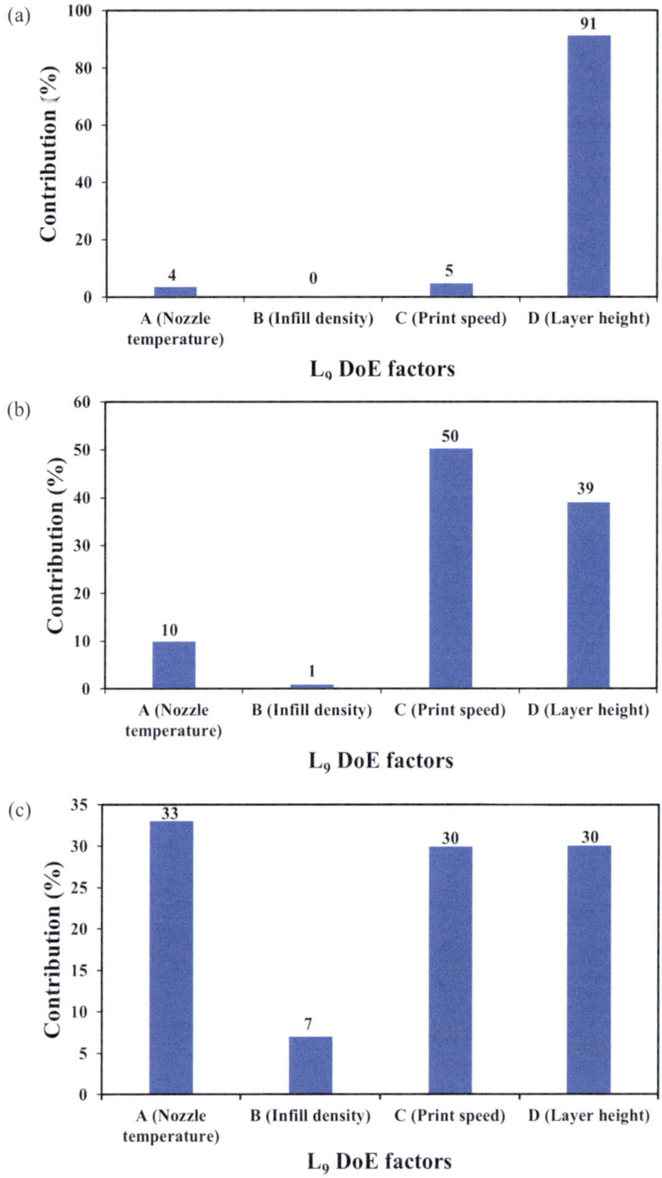

Fig. 7.5 DoE factorial contribution diagrams for the deviations of **a** thickness, **b** grip section width and **c** full length

7.3 Circularity and Cylindricity

The accuracies of circularity and cylindricity for 3D printed samples were determined using CMM-Sheffield Discovery II based on the least-square method run by PC-DMIS software. Detail elucidation with respect to circularity and cylindricity measurements, as well as measurement points was presented in Sect. 6.3.1. The measurement data of circularity and cylindricity using a radius of CAD model as the benchmark was illustrated in Fig. 7.6.

Cylindricity errors for all factor-level combinations were much greater than those of circularity. It signified that the influence associated with facet deviation along with staircase had negligible effect when compared with the shrinkage on each layer. The shrinkage caused a layer to possess distinct length sizes from the other layers, which was clearly shown by the contour of sample surfaces in the vertical (thickness) direction with a curve formation according to Fig. 7.7.

Layer height induced an important influence on circularity accuracy of 3D printed samples illustrated in Fig. 7.8a. The minimum circularity deviation was reached at the high level of layer height of 0.3 mm. It could be associated with layer number where the greater layer thickness reduced layer number and improved dimensional accuracy [8]. Nevertheless, Pandey et al. [11] evaluated a slicing method in layered manufacturing, which reported that lower layer height yielded 3D printed structures with less dimensional errors.

Another important factor that influenced circularity accuracy was print speed. Increasing print speed could reduce manufacturing time to construct a structure that decreased the shrinkage level and minimised circularity deviations. This finding was in good agreement with Žarko et al. [9] for manufacturing the samples with less dimensional error at a higher print speed. Nonetheless, Lanzotti et al. [12] reported that minimum dimensional deviation took place at the lowest level of print speed based on full factorial DoE method. Nozzle temperature was deemed a critical factor

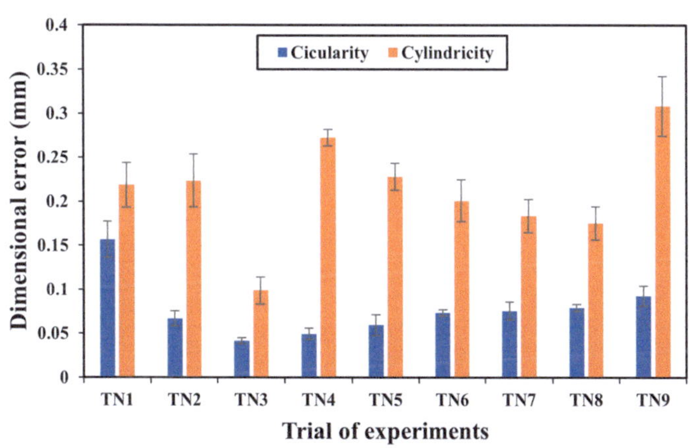

Fig. 7.6 Average dimensional deviations for circularity and cylindricity

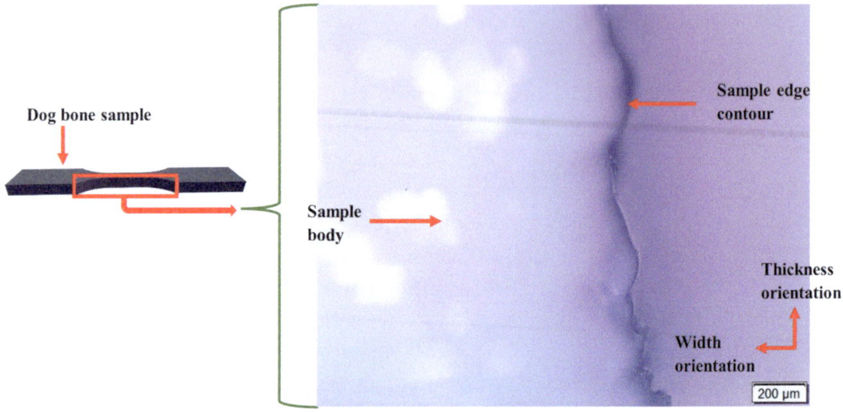

Fig. 7.7 Sample edge contour in the vertical (thickness) direction

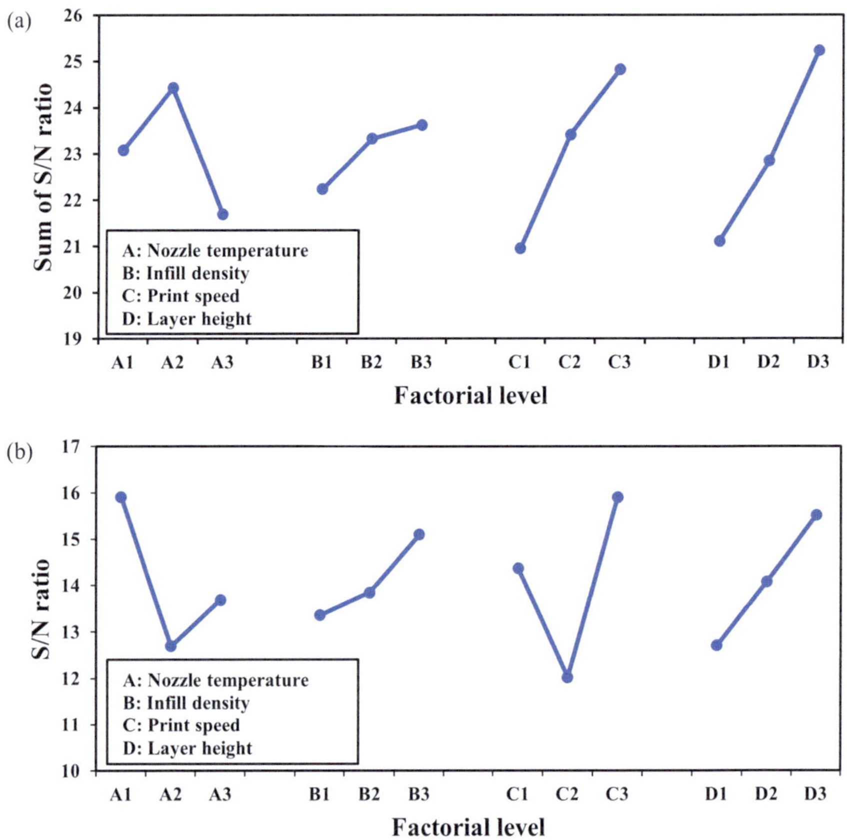

Fig. 7.8 Sum of *S/N* ratio diagram for different dimensional responses of 3D printed samples: **a** circularity and **b** cylindricity

despite its less significant effect when compared with print speed and layer height. The optimum nozzle temperature was reached at 220 °C. With increasing nozzle temperature, shrinkage effect became more pronounced, which resulted in enhancing the deviation of circularity. This was also reported by Alafaghani et al. [6] that dimensional deviations of 3D printed structures tended to be much higher with increasing nozzle temperature.

Print speed made a significant contribution to cylindricity errors of 3D printed structures according to Fig. 7.8b. The lowest cylindricity deviation was achieved at the print speed of 12 mm/s, below which cylindricity accuracy declined instead. This phenomenon could be associated with printing time and shrinkage effect in good agreement with Žarko et al. [9]. The second most significant factor was printing temperature in relation to shrinkage effect and material flow of filaments. The low level of nozzle temperature at 210 °C resulted in lower shrinkage with better dimensional accuracy. In contrast, more pronounced shrinkage effect and worse material flow of filaments arose from a relatively high level of nozzle temperature leading to greater cylindricity deviation. Such experimental results coincided with Alafaghani et al. [6] that lower extrusion temperature could reduce dimensional errors of 3D printed structures. The third most significant factor was layer height. It was manifested that a higher layer height decreased layer number with increasing cylindricity accuracy. In particular, Plaza et al. [13] evaluated the effect of layer thickness, feed rate and build direction on the geometries of 3D printed samples, resulting in lower dimensional errors owing to smaller layer thickness. On the other hand, infill density induced trivial impact on cylindricity deviations, which was deemed a non-significant factor.

Circularity of 3D printed samples could be affected by three significant factors consisting of layer height, print speed and nozzle temperature with corresponding contribution percentages of 40%, 37% and 18% respectively, as illustrated in Fig. 7.9a. Overall, optimum factor-level combination of circularity was achieved with the layer height of 0.3 mm, print speed of 12 mm/s, infill density of 100%, as well as nozzle temperature of 220 °C (i.e. $A_2B_3C_3D_3$).

In a similar manner, cylindricity was also influenced by three significant factors, namely print speed, nozzle temperature and layer height with corresponding contribution percentages of 41%, 29% and 21% shown in Fig. 7.9b. The lowest cylindricity error was obtained with the print speed of 12 mm/s, the nozzle temperature of 210 °C, the layer height of 0.3 mm and the infill density of 100% (i.e. $A_1B_3C_3D_3$).

7.3.1 Surface Finish

In general, surface quality can be associated with surface roughness and waviness of 3D printed samples. The greatest average surface roughness R_a was achieved with the nozzle temperature of 230 °C, the infill density of 20%, the print speed of 12 mm/s and the layer height of 0.2 mm (i.e. TN7). Conversely, the lowest surface roughness was obtained with the nozzle temperature of 220 °C, the infill density of 100%, the

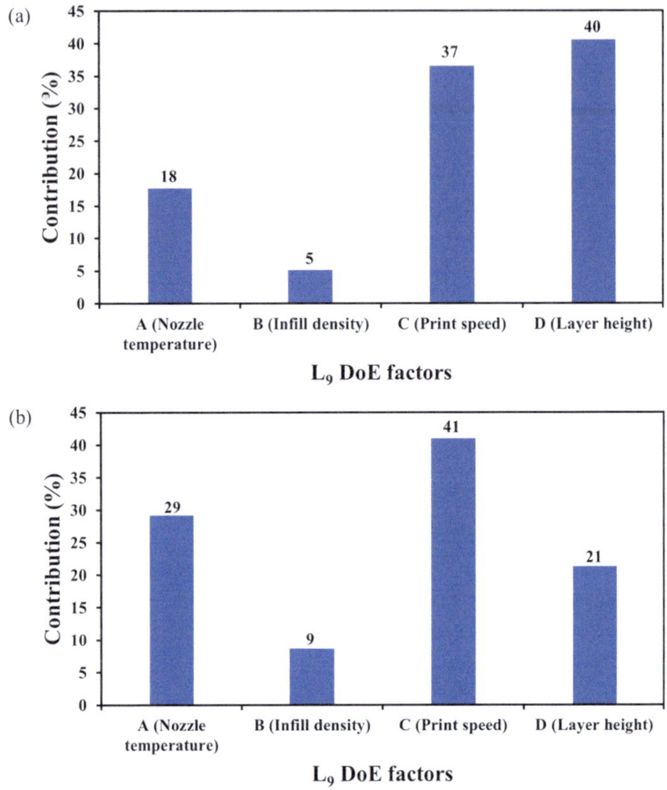

Fig. 7.9 Factorial contributions of DoE responses for 3D printed samples: **a** circularity and **b** cylindricity

print speed of 4 mm/s and the layer height of 0.2 mm (i.e. TN6), while taking into account confirmation test results illustrated in Fig. 7.10a.

In comparison, average surface waviness (W_a) of 3D printed samples was depicted in Fig. 7.10b. Standard deviations for R_a and W_a of 3D printed samples were represented by their error bars. It was worth noting that the measurements of R_a and W_a were obtained at the same regions as the data set. Better surface quality was pronounced for 3D printed samples at the nozzle temperature of 220 °C, as evidenced by R_a values of TN5, TN6 and confirmation tests. At this temperature level, supply energy was adequate to purge hot filaments out of the nozzle tip to create a smooth material layer on the print bed. When increasing the nozzle temperature up to 230 °C in case of TN7 and TN8, much rougher surfaces on samples took place, which signified a very high temperature setting to induce an overheat issue of filaments. Accordingly, the flow rate of ejected materials may not be accurately controlled on the print bed, thus deteriorating surface quality of 3D printed samples. Conversely, the lowest difference in surface roughness appeared in TN8 with a clear sign of more uniform surface quality, as illustrated by error bars in Fig. 7.10a.

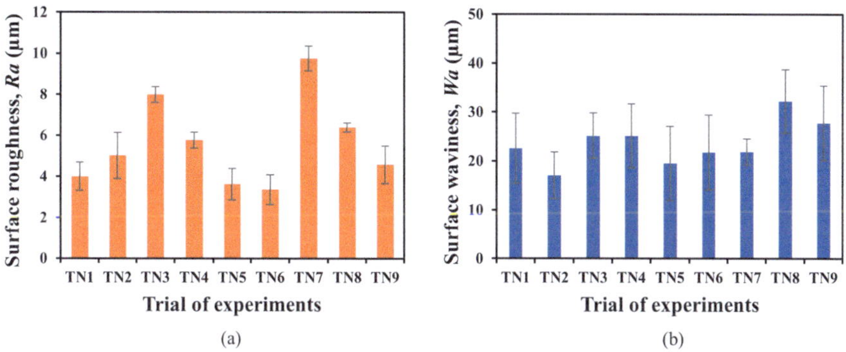

Fig. 7.10 Surface quality of 3D printed samples: **a** surface roughness and **b** surface waviness

Surface waviness of 3D printed samples ranged from 17.02 μm (i.e. TN2) to 32.21 μm (i.e. TN8). It was suggested that a 3D printed sample with the lowest R_a did not achieve the minimum W_a simultaneously. For instance, the minimum variation of W_a, represented by error bars in Fig. 7.10b, was correlated with the highest R_a for TN7. Additionally, TN9 yielded more smooth surfaces than TN7 despite a great variation with respect to surface waviness.

A 3D printing process, especially FDM, yields unequal surface quality represented by surface roughness and waviness for different sample sides consisting of top, left and right sides, as indicated in Fig. 7.11. The best surface quality was achieved on the top side of 3D printed samples according to Fig. 7.11a, which may be associated with material deposition time to build the layers. After a layer was formed on the top side of sample surface in the horizontal orientation, the nozzle then immediately deposited hot filaments on the side of that particular layer. On the other hand, it took a longer time to deposit materials above the previous layer on the left and right sides of sample surfaces along the vertical direction in order to construct a solid structure. Accordingly, adjoining layers in a horizontal direction intended to join better than those in the vertical direction in order to improve surface quality of 3D printed samples.

Average surface waviness of 3D printed samples on the top side was also lower than the other sides apart from TN3 shown in Fig. 7.11b. This exception was induced by local shrinkage to enhance surface waviness on the top side. Overall, there was a greater variation in W_a when compared with R_a, which implied that printing parameters had more significant effect on surface waviness when compared with surface roughness.

Layer height had significant influence on surface roughness of 3D printed samples shown in Fig. 7.12a, as opposed to other printing parameters. Indeed, a lower layer height decreased surface roughness [14] and yielded smoother contour on surface structures. Gauge section surfaces of 3D printed samples were scanned using Infinite Focus Alicona to examine their surface quality. As expected, a smoother contour of samples gave rise to better surface quality. Nozzle temperature induced the second

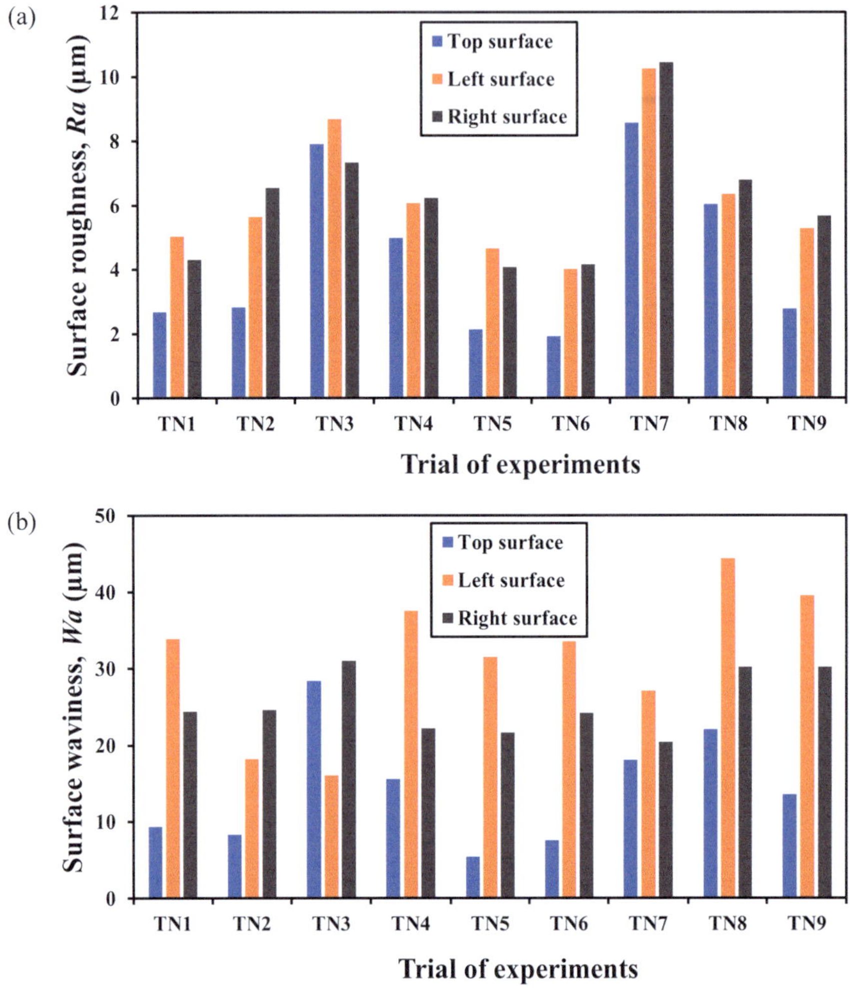

Fig. 7.11 **a** surface roughness and **b** surface waviness of 3D the samples on the top, left and right sides

significant effect on surface roughness. A higher nozzle temperature led to better surface quality when compared with layer height, which may be ascribed to material flow of deposited filaments on a print bed. The energy required to melt the filaments in the liquefier, as well as through the nozzle could be insufficient below the nozzle temperature of 220 °C. As such, it diminished material flow of deposited filaments, and thus degraded surface quality of 3D printed samples. When increasing nozzle temperature up to 230 °C, surface quality deteriorated significantly, which could be due to an overheating issue on the filaments to make them flow out of nozzle control leading to worse surface quality of 3D printed samples in good agreement with Alafaghani et al. [6].

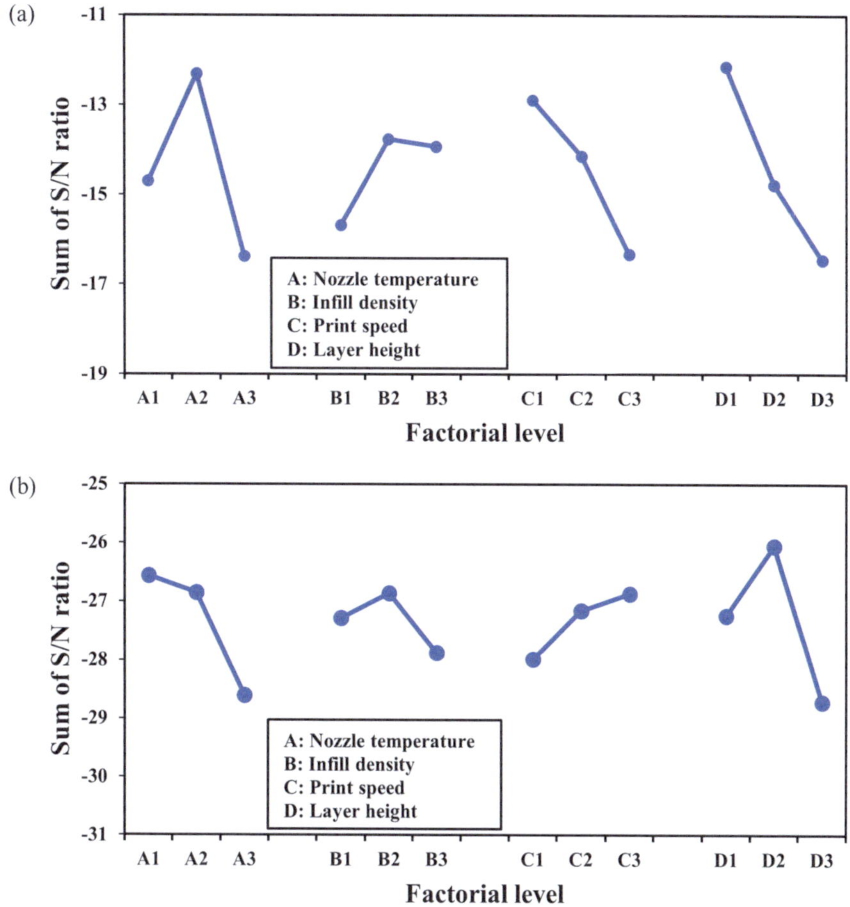

Fig. 7.12 Sum of *S/N* ratio diagram for **a** surface roughness and **b** surface waviness of 3D printed samples

Print speed was the third most significant factor to impact surface roughness of 3D printed samples. Surface roughness appeared to be enhanced with a higher print speed to decrease heated material flow when ejected via a nozzle tip. Whereas infill density was deemed a non-significant factor with trivial effect. Such a finding was in good agreement with Reddy et al. [15] when investigating the effect of layer thickness, infill density, build inclination and print quality on surface roughness of 3D printed samples.

As seen from Fig. 7.12b, layer height made the most significant contribution to surface waviness of 3D printed samples, as opposed to other printing parameters. Increasing layer height appeared to enhance surface waviness owing to shrinkage effect on sample surfaces. Overall, the minimum surface waviness could be achieved at the layer height of 0.2 mm as optimum parameter setting. Nozzle temperature became the second significant factor to influence surface waviness of 3D printed samples. Higher nozzle temperature induced more pronounced temperature gradient in between adjacent sections leading to greater surface waviness. Print speed was deemed the third significant factor, and higher print speed diminished surface waviness in relation to deposited filaments in generated layer structures. Higher print speed was detected to reduce the variation of layer thickness and further minimise surface waviness. Nonetheless, infill density had very minor effect on surface waviness, which could be considered as a non-significant factor.

Factorial contributions derived from layer height, nozzle temperature, print speed and infill density on surface roughness were determined to be 36%, 32%, 23% and 9% according to Fig. 7.13a. The optimum condition comprised the nozzle temperature of 220 °C, the infill density of 60%, the print speed of 4 mm/s, as well as the layer height of 0.1 mm (i.e. $A_2B_2C_1D_1$). As for surface waviness, the optimum condition was referred to as $A_1B_2C_3D_2$ including the nozzle temperature of 210 °C, the infill density of 60%, the print speed of 12 mm/s and the layer height of 0.2 mm shown in Fig. 7.13b.

Surface quality of FDM based structures can be improved using chemical vapour treatment, as exemplified by Lavecchia et al. [16] to make it smooth for PLA sample surfaces. PLA printed structures typically possess high surface roughness owing to staircase impact, material deposition and a slice-to-slice formation texture. Lavecchia et al. [16] utilised ethyl acetate vapours to overcome this problem in order to create smooth sample surfaces since such a solvent has favourable material merits including low cost, low toxicity and easy availability. This yielded a decrease in surface roughness by approximately 90%. Panda et al. [17] reported better surface quality of PLA structures via FDM under a chemical-vapour smoothing process using dichloromethane. It was concluded that such a chemical treatment successfully dissolved layer lines to reduce surface roughness.

Dimethyl ketone at the concentration of 99% was used for cold vapour treatment in order to improve surface quality of ABS structures via FDM [18]. Such a treatment changed the exposed surfaces by cold vapour into a soft and mushy-like condition due

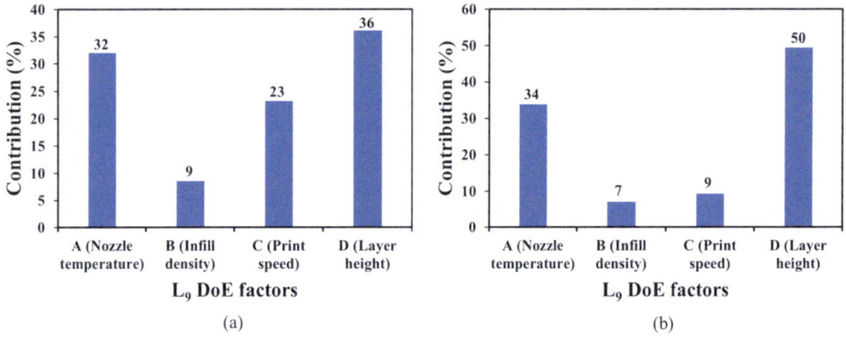

Fig. 7.13 Factorial contributions to **a** surface roughness and **b** surface waviness of 3D printed samples

to weak secondary bonds. Hence, minor material flow from polymeric layers filled the cavity region between adjacent layers, thus resulting in lower surface roughness of 3D printed structures after treatment. Pietanza et al. [19] also employed a similar material method to minimise surface roughness of FDM based structures with decreasing surface roughness R_a to 54%.

Singh et al. [20] implemented a vapour smoothing method to yield better surface quality of FDM based ABS structures. This study used highly volatile acetone that allowed for the initiation of the evaporation in a temperature range of 17–20 °C in order to meet the standard requirement of vapour smoothing station. This treatment improved the surface quality of 3D printed samples, thus influencing their dimensional features. Moreover, the use of acetone could effectively reduce the processing cost of vapour smoothing station. Chemical vapour treatment could be implemented to decrease the surface roughness of 3D printed structures made from other polymers like PU. However, an initial work would be required to select a desired solvent in order to meet the requirements for PU processing.

Triangular form was considered in infill type for 3D printing with an infill overlap of 15% and a bottom/top thickness of 1.2 mm. Figure 7.14 illustrated line patterns only in the last layer produced via FDM. Different layer thicknesses could make two distinct tilts of line patterns with the angles of 45° and 135°. Based on L_9 OA, TN1, TN5 and TN9 possessed the same angle of 135° when compared with 45° on last layers in the other types. Meanwhile, average thicknesses for 3D printed samples, built with three distinct height settings of 0.1 mm (TN1, TN5, TN9 and confirmation tests), 0.2 mm (TN2, TN6 and TN7) and 0.3 mm (TN3, TN4 and TN8), were 2.88, 2.76 and 2.94 mm respectively. Subsequently, measured average thicknesses, divided by predetermined layer heights of 0.1, 0.2 and 0.3 mm, yielded layer numbers of 29, 14 and 10 accordingly. It was manifested that 3D printed samples with an odd number of layers had a line pattern with the tilt angle of 135° on the top surface, as opposed to the others with the same angle of 45°.

7.4 Confirmation Tests

Confirmation tests were carried out to prove the optimum factor-level combination in DoE. The associated results of confirmation tests for all output parameters comprised full length, thickness, grip section width, circularity, cylindricity, surface roughness, as well as surface waviness shown in Table 7.3.

It was worth mentioning that layer height induced the most significant impact on thickness, circularity, surface roughness and surface waviness though less influential to the cylindricity. Meanwhile, print speed notably affected the grip section width and the cylindricity of 3D printed samples considering its insignificant impact on their length and surface roughness. On the other hand, nozzle temperature induced the greatest influence only on sample length, but very minor effect on its grip section width, thickness and circularity.

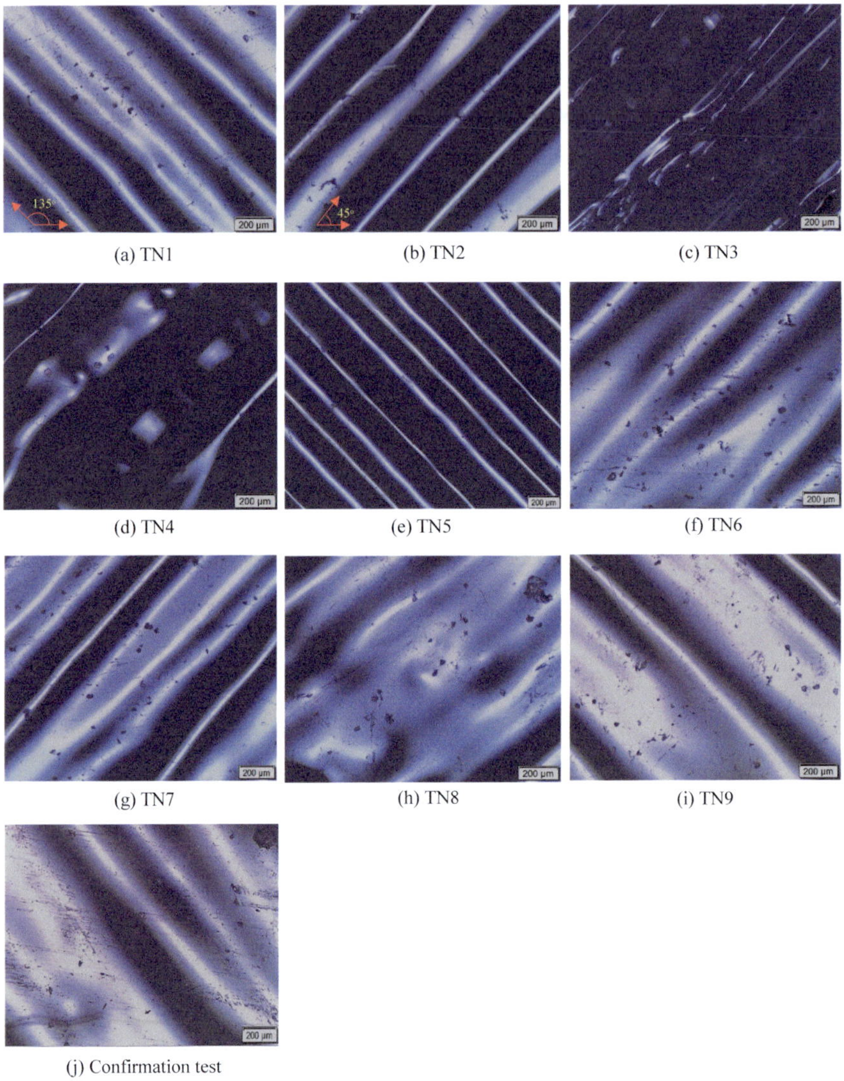

Fig. 7.14 Surface images of 3D printed samples. *Note* **a**–**i** in DoE obtained from the top side of grip section in a sequence of TN1-TN9 and **j** from confirmation tests. TN denotes the test number in DoE

All sample dimensions possessed three significant factors, consisting of nozzle temperature, layer height and print speed though thickness was only influenced by layer height. On the other hand, infill density had no critical role in affecting both dimensions and surface quality of 3D printed samples. As such, this printing parameter could be considered as a non-significant factor.

Table 7.3 Summary for DoE optimum factor-level combinations in comparison with confirmation test data

Smaller-the-better L_9 DoE response	Optimum combination	Factor-level combination	Confirmation test data
Thickness error (%)	$A_2B_1C_2D_3$	Nozzle temperature: 220 °C, infill density: 20%, print speed: 8 mm/s, layer height: 0.3 mm	6.776
Grip section width error (%)	$A_3B_3C_2D_1$	Nozzle temperature: 230 °C, infill density: 100%, print speed: 8 mm/s, layer height: 0.1 mm	0.041
Full length error (%)	$A_3B_2C_2D_3$	Nozzle temperature: 230 °C, infill density: 60%, print speed: 8 mm/s, layer height: 0.3 mm	0.119
Circularity (mm)	$A_2B_3C_3D_3$	Nozzle temperature: 220 °C, infill density: 100%, print speed: 12 mm/s, layer height: 0.3 mm	0.039
Cylindricity (mm)	$A_1B_3C_3D_3$	Nozzle temperature: 210 °C, infill density: 100%, print speed: 12 mm/s, layer height: 0.3 mm	0.099
Surface roughness (μm)	$A_2B_2C_1D_1$	Nozzle temperature: 220 °C, infill density: 60%, print speed: 4 mm/s, layer height: 0.1 mm	2.528
Surface waviness (μm)	$A_1B_2C_3D_2$	Nozzle temperature: 210 °C, infill density: 60%, print speed: 12 mm/s, layer height: 0.2 mm	13.305

The optimum factor-level combination for all sample dimensions was summarised in Table 7.3. Their significant factors were used to confirm DoE work, which aimed to assess the reproducibility of experimental data. The results of confirmation tests revealed that preferred optimum factor-level combinations for grip section width, thickness and cylindricity were referred to as TN9, TN4 and TN3. Hence, their *S/N* ratios were taken as $S/N_{confirmation}$ values according to Table 7.4. In addition, confirmation test results with respect to full-length, circularity, surface roughness and surface waviness of 3D printed samples to achieve optimum factor-level combination in Table 7.3 were 0.119%, 0.039 mm, 2.528 μm and 13.305 μm respectively. It was suggested that 3D printed samples using the optimum factor-level combination resulted in better quality than those produced from L_9 OA alone. Therefore, *S/N* ratios with respect to preferred factor-level combinations in terms of full length, circularity and surface roughness were used as $S/N_{\text{confirmation}}$ indicated in Table 7.4.

Table 7.4 presented $S/N_{\text{predicted}}$, $CI_{\text{predicted}}$, and $CI_{\text{confirmation}}$ data calculated according to Eqs. (6.6)–(6.8). The intervals of $S/N_{\text{predicted}}$ and $CI_{\text{predicted}}$ in *CI* columns were compared with $S/N_{\text{confirmation}}$ and $CI_{\text{confirmation}}$. Such intervals were calculated based on specific ranges obtained from *S/N*−CI and *S/N* + CI. The resulting data implied that the confidence intervals of prediction and confirmation tests for all sample dimensions had excellent overlapping effects to achieve good reproducibility in experimental work.

Table 7.4 Data summary for confirmation tests with respect to *S/N* ratio and confidence interval (*CI*)

Dimension	Prediction		Confirmation		Confidence interval (CI)
	$S/N_{\text{predicted}}$	CI predicted	$S/N_{\text{confirmation}}$	$CI_{\text{confirmation}}$	
Thickness	–17.581	± 1.040	–16.619	± 2.080	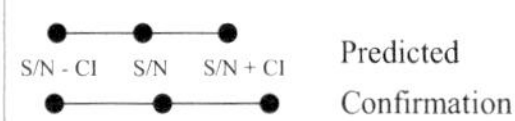
Grip section width	26.481	± 7.567	27.724	± 11.441	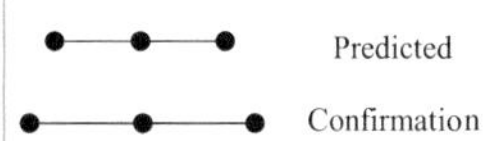
Full length	16.183	± 9.840	18.516	± 14.877	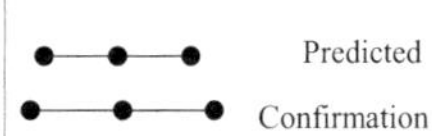
Circularity	28.349	± 4.829	28.253	± 7.301	
Cylindricity	19.120	± 5.886	20.117	± 8.899	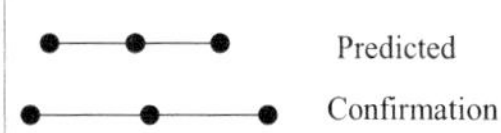
Surface roughness	–8.437	± 6.971	–8.055	± 10.536	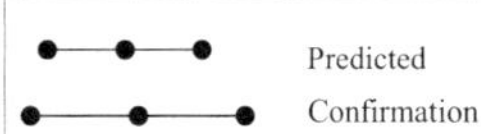
Surface waviness	–24.815	± 3.350	–22.480	± 5.065	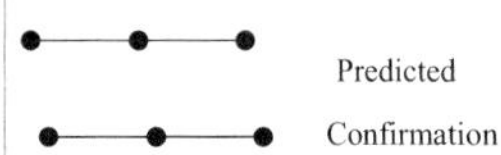

7.5 Summary

Dog-bone samples made from PU according to ASTM D638 (Type V) standard were optimally 3D printed using FDM by varying four printing parameters, consisting of infill density, nozzle temperature, print speed and layer height. The responses in this study were full length, thickness, grip section width, circularity and cylindricity of 3D printed samples, including their surface roughness and surface waviness. The shrinkage took place in all orientations, namely longitudinal (i.e. full length), transversal (i.e. grip section width) and vertical (i.e. thickness) directions. However, the shrinkage in the vertical direction induced the most significant impact when compared with other orientations.

Layer height caused the greatest influence on dimensional accuracy and surface quality of 3D printed samples when compared with other printing parameters. It had dominant effect on thickness, circularity, surface roughness, as well as surface waviness. Print speed had the greatest deviation in grip section width and cylindricity. On the other hand, nozzle temperature made the most significant contribution to full length of 3D printed samples. Nonetheless, infill density had negligible effect on all sample dimensions. The results of confirmation tests proved that experimental data were highly reproducible. In addition, research methodology used in this study could expand to other geometrical features obtained by various 3D printable polymeric materials in addition to standard dog-bone samples.

References

1. Ross, P.: Taguchi techniques for quality engineering, loss function, orthogonal experiments, parameter and tolerance design. McGraw-Hill, New York (1988)
2. Chohan, J., Singh, R., Boparai, K., Penna, R., Fraternali, F.: Dimensional accuracy analysis of coupled fused deposition modeling and vapour smoothing operations for biomedical applications. Compos. B Eng. **117**, 138–149 (2017)
3. Vasquez, J., Yepes, C., Salas, A., Solana, J., Camacho, D.: Evaluation of the metrological reliability of a graduated cylinder from experimental data from an in-situ calibration. Data Brief **32**, 106133 (2020)
4. Hanon, M., Zsidai, L., Ma, Q.: Accuracy investigation of 3D printed PLA with various process parameters and different colors. Mat. Today: Proc. **42**, 3089–3096 (2021)
5. Salkuti, P.: Quality and performance evaluation of TPU based fused deposition modeling-investigations on dimension, surface topography and mechanical properties. Master's thesis, Halmstad University (2018)
6. Alafaghani, A., Qattawi, A., Alrawi, B., Guzman, A.: Experimental optimization of fused deposition modelling processing parameters: A design-for-manufacturing approach. Procedia Manuf. **10**, 791–803 (2017)
7. Favero, C., English, J., Cozad, B., Wirthlin, J., Short, M., Kasper, F.: Effect of print layer height and printer type on the accuracy of 3-dimensional printed orthodontic models. Am. J. Orthod. Dentofac. Orthop. **152**, 557–565 (2017)
8. Zhang, C., Li, P., Chu, F., Shen, G.: Influence of the three-dimensional printing technique and printing layer thickness on model accuracy. J. Orofac. Orthop. **80**, 194–204 (2019)
9. Žarko, J., Vladic, G., Pal, M., Dedijer, S.: Influence of printing speed on production of embossing tools using FDM 3D printing technology. J. Graph. Eng. Design **8**, 19–27 (2017)

10. Gomez-Gras, G., Perez, M., Moreno, J., Pozo, G.: Experimental study on the accuracy and surface quality of printed versus machined holes in PEI Ultem 9085 FDM specimens. Rapid Prototyp. J. **27**, 1–12 (2021)
11. Pandey, P., Nallagundla, V., Dhande, S.: Slicing procedure in layered manufacturing: A review. Rapid Prototyp. J. **9**, 274–288 (2003)
12. Lanzotti, A., Martorelli, M., Staiano, G.: Understanding process parameter effects of RepRap open-source three-dimensional printers through a Design of Experiments approach. J. Manuf. Sci. Eng. **137**, 011017 (2015)
13. Plaza, E., Lopez, P., Torija, M., Muñoz, J.: Analysis of PLA geometric properties processed by FFF additive manufacturing: Effects of process parameters and plate-extruder precision motion. Polymers **11**, 1581 (2019)
14. Pandey, P., Nallagundla, V., Dhande, S.: Real time adaptive slicing for fused deposition modelling. Int. J. Mach. Tools Manuf **43**, 61–71 (2003)
15. Reddy, V., Flys, O., Chaparala, A., Berrimi, C., Krishna, A., Rosen, B.: Study on surface texture of fused deposition modeling. Procedia Manuf. **25**, 389–396 (2018)
16. Lavecchia, F., Guerra, M., Galantucci, L.: Chemical vapor treatment to improve surface finish of 3D printed polylactic acid (PLA) parts realized by fused filament fabrication. Progress Add. Manuf. **7**, 65–75 (2022)
17. Panda, S., Chabra, R., Kapil, S., Patel, V.: Chemical vapour treatment for enhancing the surface finish of PLA object producedby fused deposition method using the Taguchi optimization method. SN App. Sci. **2**, 916 (2020)
18. Garg, A., Bhattacharya, A., Batish, A.: Chemical vapor treatment of ABS parts built by FDM: Analysis of surface finish and mechanical strength. Int. J. Adv. Manuf. Techn. **89**, 2175–2191 (2017)
19. Pietanza, S., Sorgente, D., Percoco, G.: Enhancing the sustainability of chemical vapour polishing of additive manufactured ABS parts using a vacuum chamber. Rapid Prototyp. J. **23**, 1043–1050 (2017)
20. Singh, R., Singh, S., Singh, I., Fabbrocino, F., Fraternali, F.: Investigation for surface finish improvement of FDM parts by vapor smoothing process. Compos. B Eng. **111**, 228–234 (2017)

Chapter 8
Optimisation of Nanofiller Content and 3D Printing Parameters of PU/HNT Nanocomposites in Response to Dimensional Accuracy and Surface Roughness

Abstract This chapter discusses the manufacture of polyurethane (PU)/HNT nanocomposites with optimum factor-level combinations via Taguchi design of experiments (DoEs) according to L_{18} orthogonal array with the neglect of factorial interaction. The discussion was focussed on dimensional deviations of gauge section width, thickness, circularity, cylindricity, as well as surface roughness of PU/HNT nanocomposite dog-bone and cylindrical samples based on ASTM D638 (Type V) and ASTM D5963 standards via fused deposition modelling (FDM).

Keywords Optimisation · Dimensional accuracy · Surface roughness · 3D printing · PU/HNT nanocomposites

8.1 Manufacture of PU/HNT Nanocomposite Films

Polyurethane (PU) in MM-4520 grade with a shape memory behaviour was supplied by SMP Technologies, Inc., Japan. It has a glass transition temperature (T_g) of 45 °C and a melt viscosity of 3310 Pa·s at 215 °C [1]. In addition, halloysite nanotubes (HNTs) as additives in powder form were provided by Imerys Ceramics, New Zealand [2]. Both material properties were listed in Tables 6.2 and 6.3 respectively. Dimethylformamide (DMF), purchased from ChemSupply, Australia, was used as an effective chemical solvent to dissolve PU pellets. All materials were selected without any further modification. PU/HNT nanocomposites were manufactured with three major steps including solution casting, filament extrusion and FDM-based 3D printing with a detailed flow chart illustrated in Fig. 6.2.

W. T. Nugroho et al., *Thermoplastic Polyurethane (TPU)/Halloysite Nanotube (HNT) Nanocomposites*, Engineering Materials, https://doi.org/10.1007/978-981-95-0541-8_8

8.2 Optimisation Techniques

The effect of HNT content and 3D printing parameters on dimensional deviation and surface quality of PU/HNT nanocomposites were evaluated by means of L_{18} Taguchi orthogonal array (OA) according to Table 8.1 with the aid of Pareto analysis of variance (ANOVA). All 3D printing parameters, comprising print speed, nozzle temperature, layer height and infill density, were set at three different levels in addition to six different levels for HNT content, as shown in Table 8.2. Minimal factorial interaction was neglected in this DoE work for simplicity. DoE responses were referred to as minimum dimensional deviation and surface roughness of PU/HNT nanocomposites using sum of signal-to-noise (*S/N*) ratios. *S/N* ratio can be defined as a critical indicator in robust design and manufacturing to improve the quality and measurements with variability reduction [3]. Accordingly, "the smaller-the-better" criterion [4] was followed using Eq. (8.1):

$$S/N_i = -10Log_{10}\frac{1}{n_i}\left(\sum_{u=1}^{n_i} y_u^2\right) \tag{8.1}$$

where i, n_i, u and y_u denote the number of experiments, the number of trials for the ith experiment, the trial number and data observed as an output response. In general, the greatest *S/N* ratio yields the maximum result, thus leading to the best quality with the lowest variance. This is a fundamental principle to follow for the selection of optimum factor-level combinations in this study.

8.3 Thickness and Gauge Section Width of Dog-Bone Samples

As mentioned in Chap. 7, dimensional accuracy of 3D printed samples was characterised by their dimensional deviation against a CAD model, which is also known as dimensional error. Figure 8.1 demonstrated calculation results with respect to absolute dimensional errors for gauge section width and thickness of 3D printed samples. Apparently, the presence of nanofillers reduced dimensional accuracy of 3D printed polymer nanocomposite samples [7]. In comparison with gauge section width, the thickness induced much greater deviation. In particular, the inclusion of 10 wt% HNTs in polymer nanocomposites (i.e. TN16) yielded an increase by 13.37% in thickness variation, as opposed to 4.67% for that of neat PU (i.e. TN1). HNT aggregates were believed to contribute partly to a lumpy characteristic in thickness with higher dimensional errors for nanocomposite samples. More clearly, much higher dimensional deviations appeared with respect to sample thickness in z vertical direction when compared with those in x and y horizontal directions for the width and the length of 3D printed nanocomposite samples due to shrinkage effect [8]. TN17 and TN2 material samples with the addition of 10 and 0 wt% HNTs gave

Table 8.1 L_{18} OA for PU/HNT nanocomposites used in this study [5]

Exp	Symbol	Factor				
		A (HNT content wt%)	B (nozzle temperature °C)	C (print speed mm/s)	D (infill density %)	E (layer height mm)
1	TN1	0	210	10	40	0.2
2	TN2	0	220	20	70	0.3
3	TN3	0	230	30	100	0.4
4	TN4	2	210	10	70	0.3
5	TN5	2	220	20	100	0.4
6	TN6	2	230	30	40	0.2
7	TN7	4	210	20	40	0.4
8	TN8	4	220	30	70	0.2
9	TN9	4	230	10	100	0.3
10	TN10	6	210	30	100	0.3
11	TN11	6	220	10	40	0.4
12	TN12	6	230	20	70	0.2
13	TN13	8	210	20	100	0.2
14	TN14	8	220	30	40	0.3
15	TN15	8	230	10	70	0.4
16	TN16	10	210	30	70	0.4
17	TN17	10	220	10	100	0.2
18	TN18	10	230	20	40	0.3

rise to maximum and minimum errors for gauge section width at 6.62% and 2.76% respectively. Finally, non-uniform HNTs, especially at high content levels, enabled to decrease dimensional accuracy of nanocomposite samples, as opposed to neat PU with much better material flow and homogeneity.

Figure 8.2a illustrated prevalent effect of HNT content and layer height on dimensional deviation of thickness for 3D printed nanocomposite samples, which was deemed significant factors with their cumulative contribution up to 95%. Conversely, print speed, infill density and nozzle temperature were considered as non-significant factors with their corresponding minor contributions of 1%, 1% and 0% accordingly. Meantime, dimensional accuracy of gauge section width could be affected by both HNT content and print speed, resulting in their cumulative contribution of 99% according to Fig. 8.2b. It was worth mentioning that the inclusion of HNTs influenced 3D printed nanocomposite samples in both vertical and horizontal directions. The effect of print speed was associated with shrinkage effect, which meant that nanocomposite samples at a low speed required much longer time for material solidification when compared with those at the relatively high speed with more noticeable shrinkage. This further increased their dimensional deviations. Finally, nozzle temperature, layer height and infill density yielded very minor effect with

Table 8.2 Fixed 3D printing parameters used for material sample fabrication [6]

Parameter	Specific parameter	Setting
Quality	Shell thickness (mm)	1.6
	Initial layer thickness (mm)	0.5
	Initial layer line width (%)	120
	Top surface quality	Precise
Fill	Bottom/top thickness (mm)	1.2
	Infill interface density	Dense
	Infill type	Triangle
	Infill overlap (%)	15
Temperature	Bed temperature (°C)	55
Speed	Travel speed (mm/s)	150
	Bottom layer speed (mm/s)	15
	Infill speed (mm/s)	30
Filament	Flow (%)	115
Retraction	Speed (mm/s)	30
	Distance (mm)	5
	Minimum travel (mm)	1.5
	Minimal extrusion before retracting (mm)	0.005

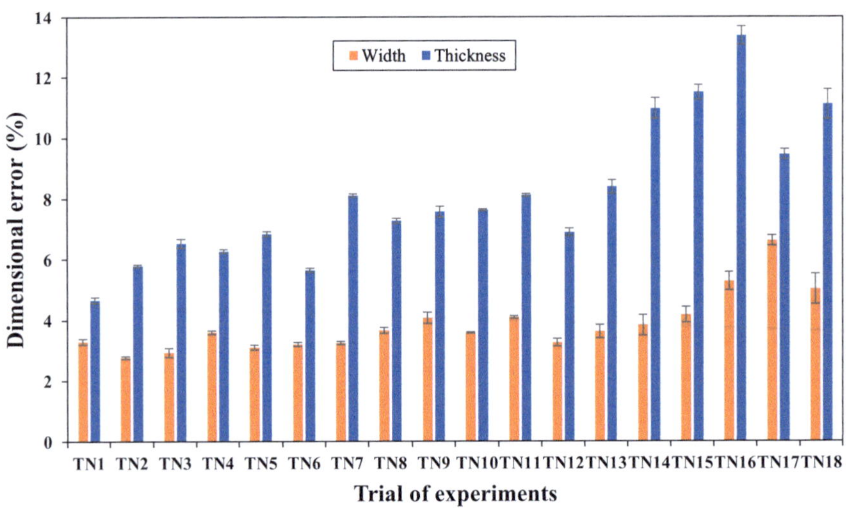

Fig. 8.1 Effects of HNT content and 3D printing parameters on dimensional deviations of gauge section width and thickness of nanocomposite samples

their corresponding contributions of 1%, 0% and 0% only, which could be deemed non-significant factors.

It was evidently observed from the sum of *S/N* ratios in Fig. 8.3a that increasing HNT content could reduce dimensional accuracy of thickness for nanocomposite samples owing to typical HNT aggregation. A similar trend was also demonstrated by layer height where the best thickness accuracy was achieved when nanocomposite samples were 3D printed using the thinnest layers [9]. The optimum factor-level combination for the thickness of 3D printed nanocomposite samples was referred to as $A_1B_1C_1D_3E_1$ with the HNT content of 0 wt%, the nozzle temperature of 210 °C, the infill density of 100%, as well as the layer height of 0.2 mm.

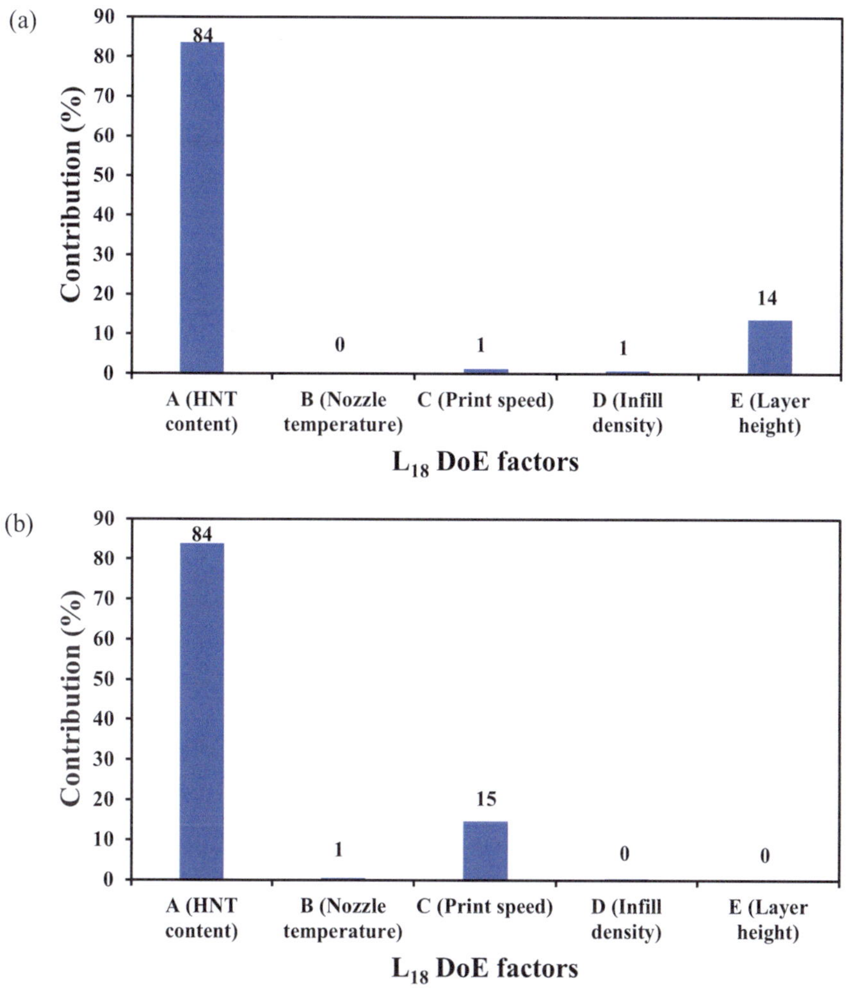

Fig. 8.2 Factorial contribution diagrams to **a** thickness and **b** gauge section width of nanocomposite samples

(a)

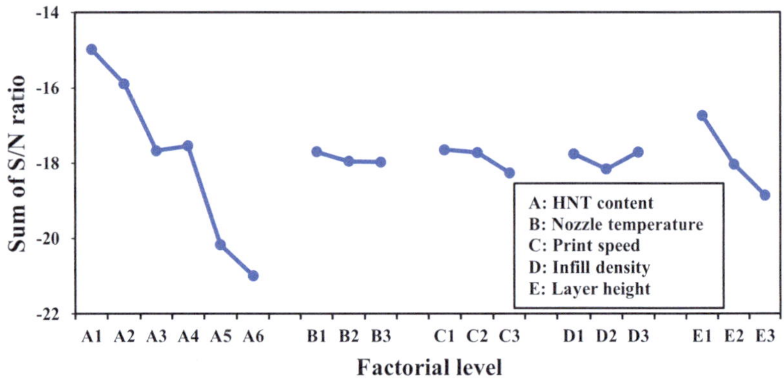

(b)

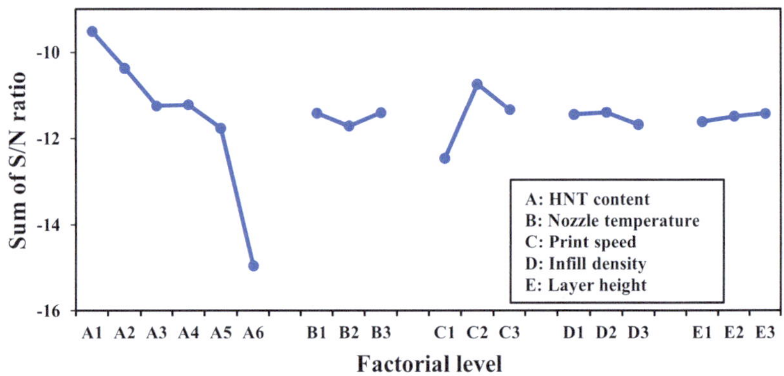

Fig. 8.3 Sum of *S/N* ratio diagrams for **a** thickness and **b** gauge section width of 3D printed nanocomposite samples

Neat PU had the lowest dimensional errors with respect to gauge section width for 3D printed nanocomposite samples. Its accuracy declined with increasing HNT content shown in Fig. 8.3b. On the contrary, a higher print speed up to 20 mm/s diminished width deviation owing to manufacturing time and shrinkage effect. Nonetheless, with a further increase in print speed to 30 mm/s, the accuracy of gauge section width slightly decreased, which might be attributed to a certain temperature gradient taking place at the threshold of print speed to influence a fluid-like behaviour of melted feedstock materials, namely the flow rate of melted materials. As such, layer width was reduced with increasing dimensional deviation of sample width. In contrast, nozzle temperature, infill density and layer height had negligible effects. Overall, optimum factor-level combination with respect to dimensional accuracy of gauge section width for 3D printed nanocomposite samples took place with the HNT content of 0 wt%, the nozzle temperature of 230 °C, the print speed of 20 mm/s, as well as the layer height of 0.3 mm (i.e. $A_1B_3C_2D_2E_3$).

8.4 Circularity, Cylindricity and Surface Roughness of Dog-Bone Samples

The measurement data concerning circularity and cylindricity of nanocomposite dog-bone samples using the least square method via PC-DMIS software on Sheffield Discovery II, a coordinate measurement machine (CMM), were presented in Fig. 8.4. The dimensional errors of cylindricity for all factor-level combinations were greater than those of circularity. It indicated that very little impact emerged from facet deviation and staircase issues, as opposed to the shrinkage effect on each material layer. The greatest deviations of circularity were reached at 0.112 and 0.18 mm respectively, in relation to TN1 for neat PU and TN16 for PU/10 wt% HNT nanocomposites. The addition of HNTs tended to diminish circularity accuracy for nanocomposite dog-bone samples by approximately 38% owing to HNT agglomeration. Besides, minimum and maximum errors of cylindricity took place for TN3 (i.e. neat PU) and TN18 (i.e. 10 wt% HNTs), resulting in corresponding cylindrical deviations of 0.192 and 0.244 mm respectively.

HNT content and layer height, represented by factors A and E, were deemed the two most significant factors in relation to circularity because of their prevalent contributions of 56 and 42% according to Fig. 8.5a. Other non-significant factors, namely nozzle temperature, infill density and print speed, made trivial impact with their corresponding contributions of 0%, 1% and 1% accordingly. On the other hand, HNT content made the highest contribution to cylindricity, which was then followed by print speed despite minor impact induced by nozzle temperature, layer height and infill density depicted in Fig. 8.5b.

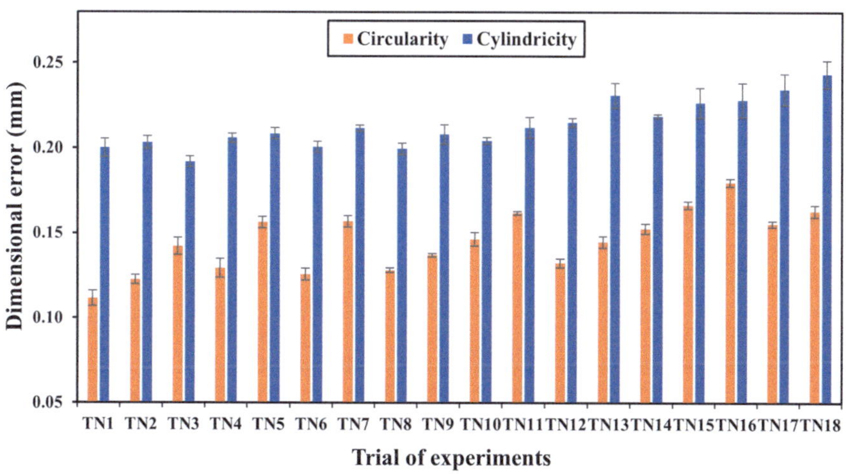

Fig. 8.4 Dimensional deviations of circularity and cylindricity of 3D printed nanocomposite dog-bone samples

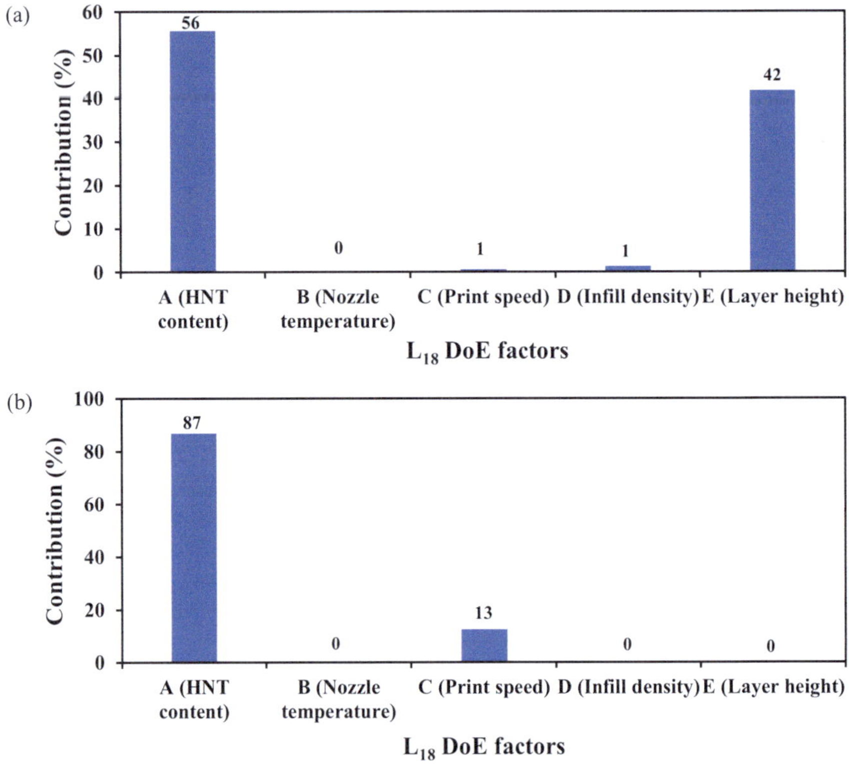

Fig. 8.5 Factorial contribution diagrams for **a** circularity and **b** cylindricity of 3D printed nanocomposite dog-bone samples

As seen in Fig. 8.6a, increasing HNT content could reduce the accuracy of circularity for 3D printed nanocomposite samples. The presence of rigid nanofillers such as HNTs in PU matrices deteriorated facet deviation and staircase effect, leading to higher dimensional errors of circularity. Accordingly, the minimum layer height of 0.2 mm resulted in the lowest shrinkage, thus yielding the minimum circularity deviation in good agreement with Plaza et al. [10]. In general, the lowest error of circularity was acquired with the HNT content of 0 wt%, the nozzle temperature of 230 °C, the print speed of 10 mm/s, the infill density of 70% and the layer height of 0.2 mm (i.e. $A_1B_3C_1D_2E_1$). Furthermore, the addition of HNTs within PU matrices also enhanced the deviation of cylindricity for 3D printed nanocomposite dog-bone samples with respect to manufacturing time and shrinkage effect, as illustrated in Fig. 8.6b. A higher print speed reduced manufacturing time and shrinkage for the improvement of dimensional accuracy in good agreement with Zarko et al. [11]. In summary, the highest accuracy with respect to cylindricity was achieved with the HNT content of 0 wt%, the nozzle temperature of 220 °C, the print speed of 30 mm/s, the infill density of 100%, as well as the layer height of 0.4 mm (i.e. $A_1B_2C_3D_3E_3$).

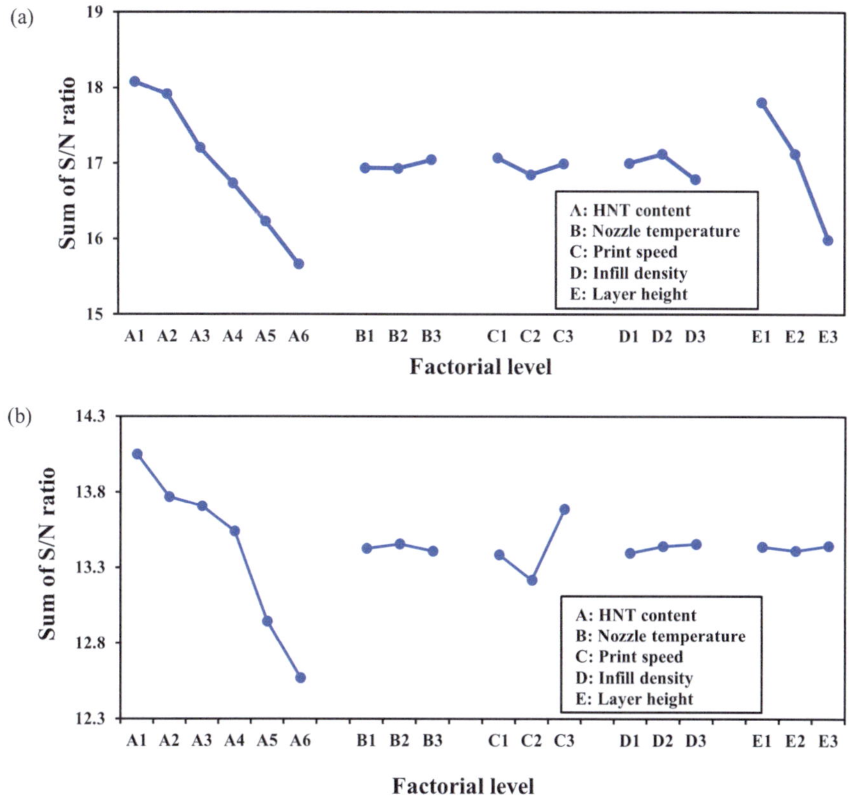

Fig. 8.6 Sum of *S/N* ratio diagrams with respect to dimensional errors of 3D printed nanocomposite dog-bone samples for **a** circularity and **b** cylindricity

Surface roughness of 3D printed nanocomposite dog-bone samples was measured on the top, the right and the left sides with an optical 3D measurement instrument Infinite Focus Alicona (Bruker Alicona, Raaba, Austria) based on a light source to scan the desired regions. Detailed information regarding measured top, right and left sides, as well as specific steps to evaluate sample surfaces were mentioned in Sect. 6.3.1. It was seen from Fig. 8.7 that the top side of sample surface achieved the best surface quality for all sample types, which may be associated with deposition time of filament materials to build up 3D printed layers. After the layer construction on the top side of sample surface in the horizontal direction, the nozzle ejected hot filaments next to the previous layer side immediately. In comparison, right and left sides of sample surfaces in the vertical direction required a much longer time for material deposition on the previous layer, thus resulting in worse surface quality and more severe shrinkage.

Consequently, adjacent layers in the horizontal direction yielded smoother surfaces when compared with those in the vertical direction. Overall, the highest

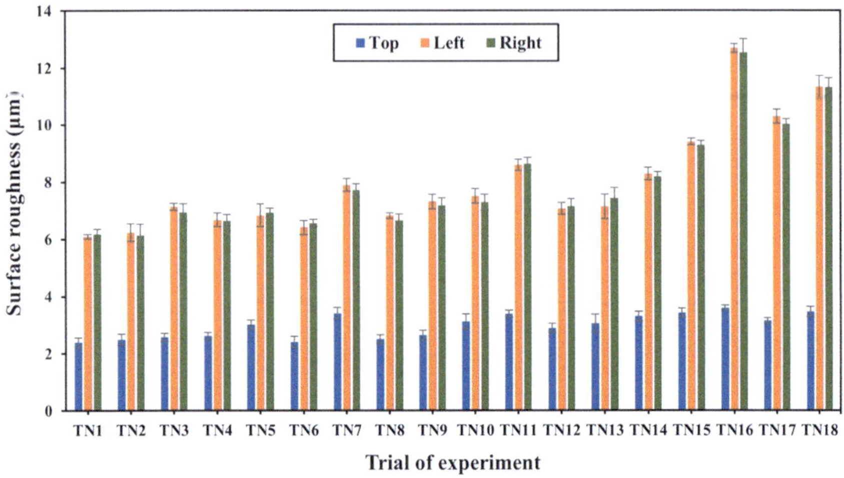

Fig. 8.7 Surface roughness of 3D printed nanocomposite dog-bone samples

surface quality was obtained without HNTs (i.e. TN1) for all factor-level combinations, which was opposed to the roughest surface observed in TN16 for 3D printed nanocomposite dog-bone samples.

Similar to relevant results obtained in relation to circularity and cylindricity, HNT content was also detected to significantly influence surface roughness of 3D printed nanocomposite dog-bone samples with the highest contribution of 85%, which was then followed by layer height at 14% according to Fig. 8.8a. As such, both HNT content and layer height were taken into account as significant factors with a cumulative contribution over 90%. Other 3D printing parameters, namely nozzle temperature, infill density and print speed only had almost no contributions at 0%, 1% and 0% respectively, which were then categorised as non-significant factors.

It was clearly seen in Fig. 8.8b that the inclusion of HNTs within PU matrices undermined surface quality of 3D printed nanocomposite dog-bone samples. As expected, increasing HNT content promoted surface roughness level for such material samples due to the existence of HNTs as rigid additives in different sizes. A much thicker layer induced a rougher surface contour of 3D printed nanocomposite samples leading to detrimental surface quality, which was in good agreement with Pandey et al. [12]. Overall, the best factor-level combination with respect to surface roughness was obtained with the HNT content of 0 wt%, the nozzle temperature of 220 °C, the infill density of 100%, the print speed of 20 mm/s, as well as the layer height of 0.2 mm (i.e. $A_1B_2C_2D_3E_1$).

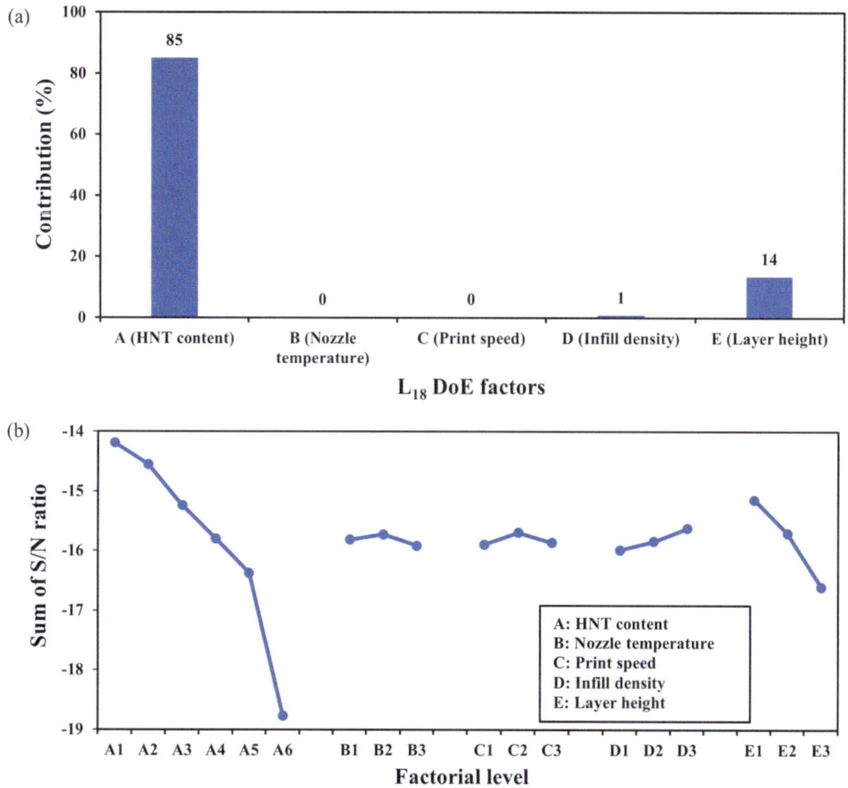

Fig. 8.8 **a** Factorial contribution diagram and **b** sum of *S/N* ratio diagram for surface roughness of 3D printed nanocomposite dog-bone samples

8.5 Thickness, Circularity, Cylindricity and Surface Quality of Cylindrical Samples

The presence of HNTs within PU matrices tended to increase thickness deviations of nanocomposite cylindrical samples shown in Fig. 8.9, arising from HNT aggregation and shrinkage effect. The lowest and highest dimensional deviations for such cylindrical samples were determined to be 2.06% and 11.023% at the HNT contents of 0 and 10 wt% with respect to TN1 and TN16 respectively. Apparently, two significant factors of HNT content and layer height made the contributions of 64% and 33% respectively with respect to thickness deviation according to Fig. 8.10a. In the meantime, the rest of non-significant factors ranging from nozzle temperature, print speed to infill density induced unnoticeable impact with very little contributions of 0%, 1% and 2% accordingly.

As depicted in Fig. 8.10b, increasing HNT content from 0 to 10 wt% with respect to A_1 and A_6 reduced the sum of *S/N* ratios, which signified a declining trend in

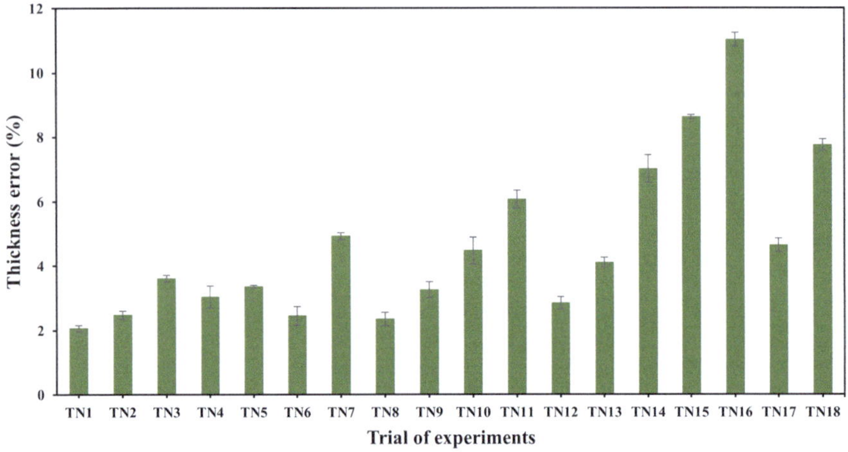

Fig. 8.9 Effects of printing parameters and HNT content on thickness errors of 3D printed nanocomposite cylindrical samples

thickness accuracy of nanocomposite cylindrical samples associated with typical HNT aggregation. Accordingly, the lowest deviation of thickness was achieved at the minimum layer height of 0.2 mm with less shrinkage. Overall, the optimum factor-level combination with respect to thickness deviation of nanocomposite cylindrical samples was obtained with the HNT content of 0 wt%, the nozzle temperature of 220 °C, the print speed of 20 mm/s, as well as the layer height of 0.2 mm (i.e. $A_1B_2C_2D_3E_1$).

In general, the deviations of cylindricity for all factor-level combinations tended to be much higher than those of circularity. It was manifested that facet deviation along with staircase effect made negligible influence when compared with shrinkage effect identified at each material layer according to Fig. 8.11. Both minimum and maximum circularity errors up to 0.170 and 0.307 mm appeared for TN1 (neat PU) and TN16 with 10 wt% HNTs, as opposed to the lowest and highest errors of cylindricity at 0.225 and 0.395 mm in relation to TN3 for neat PU and TN18 with 10 wt% HNTs.

It was seen in Fig. 8.12a that HNT content and layer height were deemed two significant factors in relation to circularity error with their contributions of 80% and 19% respectively. As earlier mentioned, facet deviation, shrinkage and staircase effect might be associated with HNT content and layer height accordingly. Meanwhile, infill density, nozzle temperature and print speed had unnoticeable influence, as evidenced by their corresponding contributions of 1%, 0% and 0%. On the other hand, HNT content and print speed achieved a cumulative contribution of approximately 97% notwithstanding their trivial impact arising from layer height, infill density and nozzle temperature with corresponding contributions of 2%, 1% and 0% shown in Fig. 8.12b.

It could be clearly seen in Fig. 8.13a that low HNT content enhanced circularity accuracy of 3D printed nanocomposite cylindrical samples. The decrease in circularity accuracy could be induced by the inclusion of HNTs as rigid nanofillers

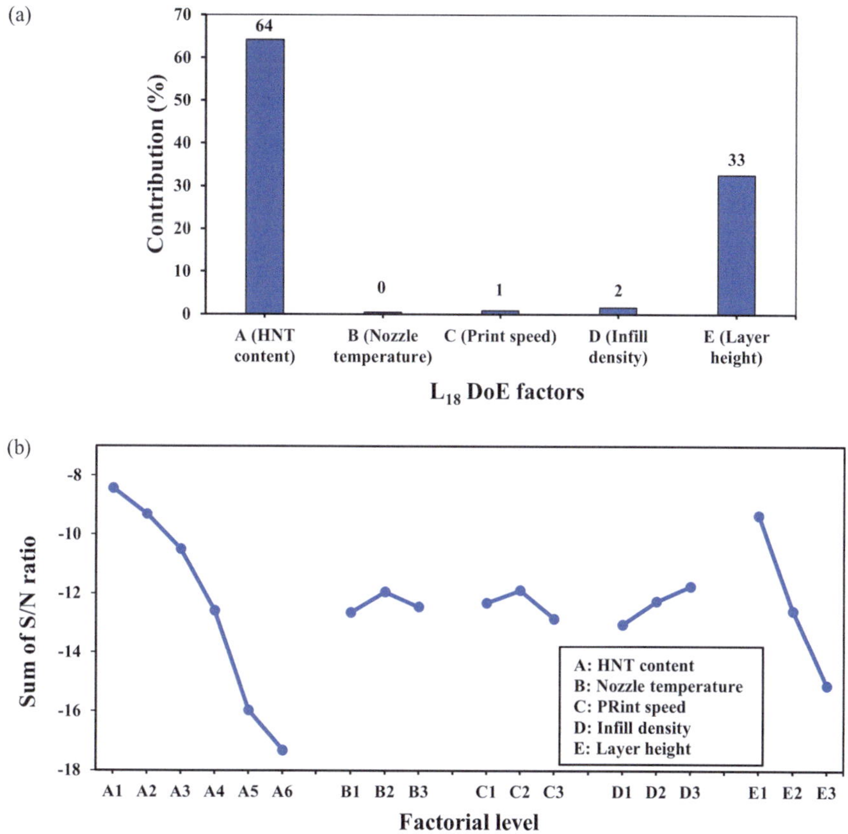

Fig. 8.10 **a** Factorial contribution diagram and **b** sum of *S/N* ratio diagram for thickness deviation of 3D printed nanocomposite cylindrical samples

to enhance facet deviation and staircase effect. Besides, layer height also played a leading role in this result where a low layer height made the minimum dimensional deviation of 3D printed structures, as mentioned by Pandey et al. [12]. This finding was contradictory to the result reported by Zhang et al. [9] that higher layer thickness minimised layer number, and further contributed to the improvement of dimensional accuracy for nanocomposite samples. On the contrary, other factors, comprised of nozzle temperature, infill density and print speed, were determined to be non-significant factors due to their minor effect on the deviation of cylindricity.

It was observed in Fig. 8.13b that accuracy of cylindricity for 3D printed cylindrical samples declined with increasing the HNT content owing to HNT aggregation at higher filler content levels. Conversely, the deviation of cylindricity diminished with increasing speed arising from manufacturing time and shrinkage effect. Apparently, a higher print speed yielded better accuracy of cylindricity due to the decrease in manufacturing time, as well as less adverse shrinkage effect.

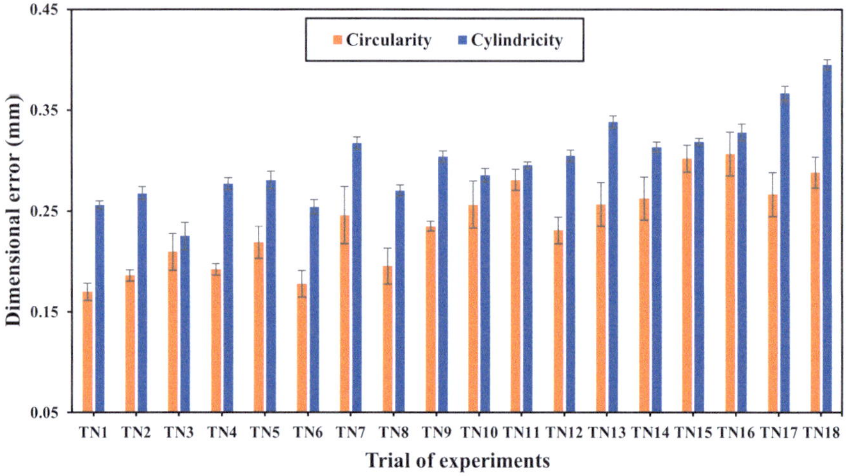

Fig. 8.11 Dimensional deviations of circularity and cylindricity for 3D printed nanocomposite cylindrical samples

Surface roughness of nanocomposite cylindrical samples was examined only on the top and side surfaces, which was slightly different from that of dog-bone counterparts. It could be ascribed to the bottom section in possession of smooth and glossy surfaces, resulting from the interaction between the first deposited filaments subjected to the load obtained from the total weight of subsequent layers and the print bed. The measurements of surface roughness for all cylindrical samples were repeated three times with detailed information available at the Sect. 6.3.1.

Figure 8.14 demonstrated that the top surface of cylindrical samples had smooth surfaces for all sample types when compared with their side surfaces, which might be attributed to the deposition time of filaments to build 3D printed layers. When a layer was deposited on the top surface of the sample in the horizontal direction, the nozzle shortly ejected and directed hot materials to the layer side. In the meantime, the sample side in the vertical direction appeared to require a much longer time to deposit hot filaments onto previous layers, further resulting in more visible shrinkage and higher surface roughness. Accordingly, adjacent layers in the horizontal direction exhibited better surface quality than those in the vertical direction. The best surface quality for the top surface was obtained for TN1 without HNTs among all factor-level combinations leading to the surface roughness of 2.23 μm, which was opposed to the worst scenario for TN16 with corresponding surface roughness of 5.64 μm. Moreover, the side part of sample surfaces achieved the smoothest surface with the roughness of 6.18 μm for TN1 relative to the roughest surface with 12.10 μm for TN16.

Generally speaking, HNT content had major effect on surface roughness of 3D printed nanocomposite samples with its contribution of approximately 84%, which was followed by layer height at 13% according to Fig. 8.15a. As such, such input

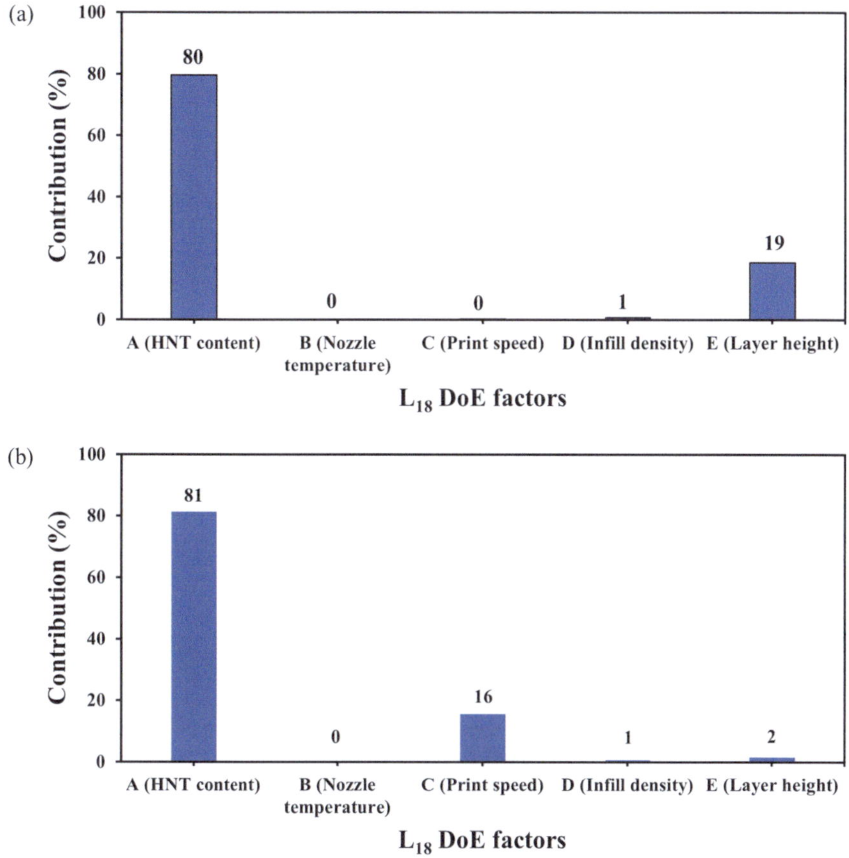

Fig. 8.12 Factorial contribution diagrams for **a** circularity and **b** cylindricity of 3D printed nanocomposite cylindrical samples

parameters were deemed two significant factors owing to their cumulative contribution exceeding 90%. On the other hand, nozzle temperature, print speed and infill density with their corresponding contributions of 1%, 1% and 1% could be classified as non-significant factors. Figure 8.15b indicated that the addition of HNTs into PU matrices undermined surface quality of 3D printed nanocomposite cylindrical samples. The best surface quality was observed at the lowest layer height. The minimum surface roughness was obtained with the HNT content of 0 wt%, the nozzle temperature of 210 °C, the print speed of 20 mm/s and the layer height of 0.2 mm (i.e. $A_1B_1C_2D_3E_1$).

Confirmation test results with respect to all dimensional deviations in terms of thickness, gauge section width, circularity, cylindricity and surface quality for both dog-bone and cylindrical samples were summarised in Table 8.3. It was manifested that HNT content and layer height resulted in significant impact on thickness, circularity and surface quality of dog-bone samples. Meanwhile, gauge section width

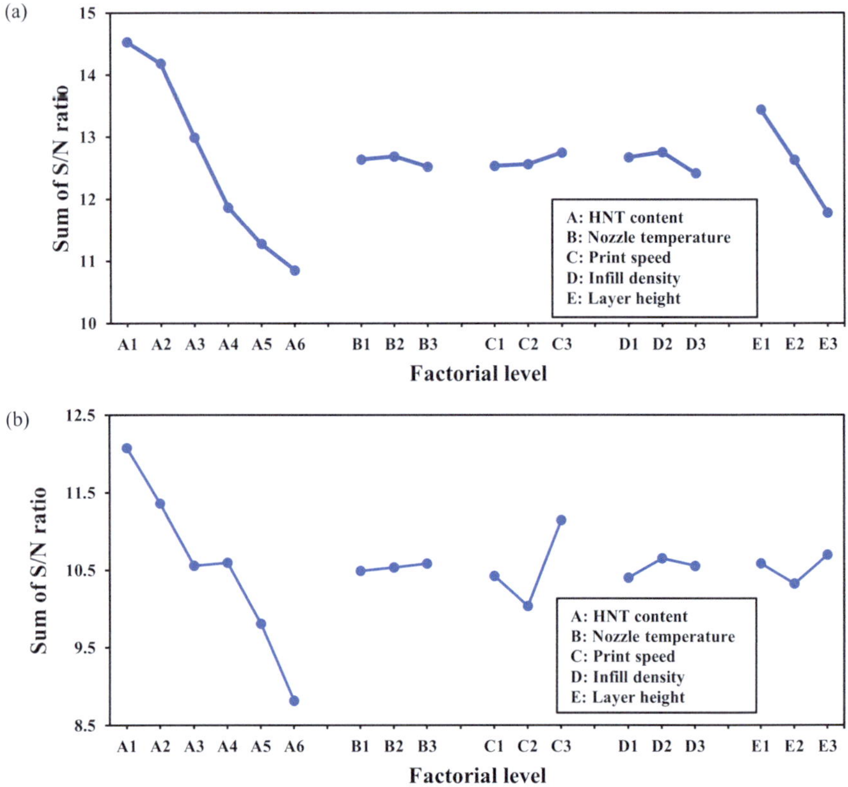

Fig. 8.13 Sum of *S/N* ratio diagrams for **a** circularity and **b** cylindricity of 3D printed nanocomposite cylindrical samples

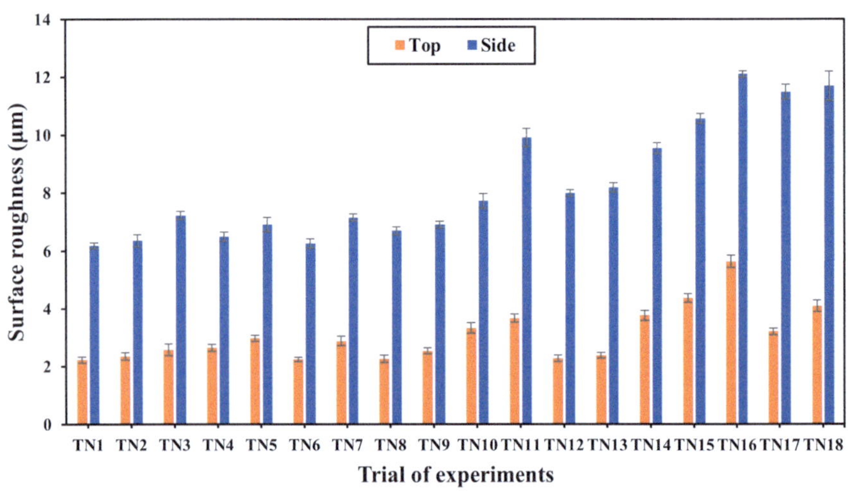

Fig. 8.14 Surface roughness of nanocomposite cylindrical samples

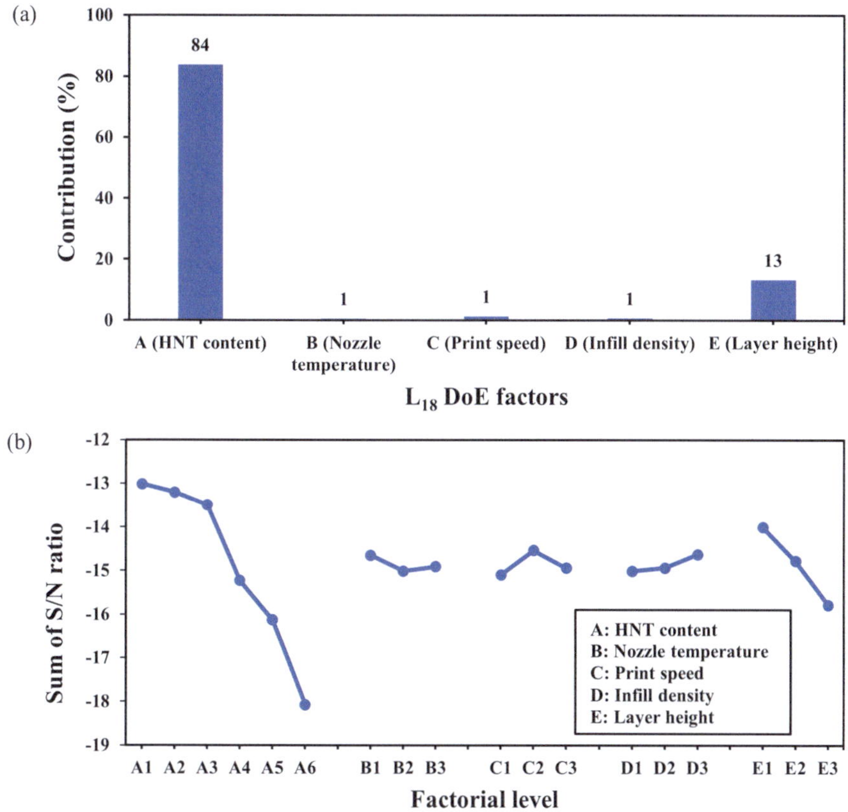

Fig. 8.15 **a** Factorial contribution diagram and **b** sum of *S/N* ratio diagram for surface roughness of 3D printed nanocomposite cylindrical samples

and cylindricity were primarily influenced by HNT content and print speed. As for cylindrical samples, thickness, circularity and surface quality were greatly influenced by HNT content and layer height. On the other hand, HNT content and print speed became key factors for the accuracy of cylindricity. Other input parameters like nozzle temperature and infill density had negligible impact for both dog-bone and cylindrical samples as non-significant factors.

Calculation results obtained from confirmation tests with respect to thickness, gauge section width, circularity, cylindricity and surface quality of dog-bone samples were found to be 4.583%, 2.673%, 0.111 mm, 0.191 mm and 4.783 μm respectively for their optimum factor-level combination according to Table 8.3. Conversely, the highest accuracies of thickness, circularity, cylindricity and surface quality for cylindrical nanocomposites samples in confirmation tests were achieved with their corresponding data of 1.970%, 0.164 mm, 0.224 mm and 4.125 μm accordingly. Nanocomposite samples based on optimum factor-level combination yielded better accuracy when compared with those obtain directly from L_{18} OA. Consequently, *S/*

Table 8.3 Summary for DoE optimum factor-level combinations and confirmation tests

Smaller-the-better L_9 DoE response	Structure	Optimum combination	Factor-level combination	Confirmation test result for dimensional errors
Thickness error (%)	Dog-bone	$A_1B_1C_1D_3E_1$	HNT level: 0 wt%, nozzle temperature: 210 °C, print speed 10 mm/s, infill density: 100%, layer height: 0.2 mm	4.583
Gauge section width error (%)	Dog-bone	$A_1B_3C_2D_2E_3$	HNT level: 0 wt%, nozzle temperature: 230 °C, print speed 20 mm/s, infill density: 70%, layer height: 0.4 mm	2.673
Circularity (mm)	Dog-bone	$A_1B_3C_1D_2E_1$	HNT level: 0 wt%, nozzle temperature: 230 °C, print speed 10 mm/s, infill density: 70%, layer height: 0.2 mm	0.111
Cylindricity (mm)	Dog-bone	$A_1B_2C_3D_3E_3$	HNT level: 0 wt%, nozzle temperature: 220 °C, print speed 30 mm/s, infill density: 100%, layer height: 0.4 mm	0.191
Surface roughness (μm)	Dog-bone	$A_1B_2C_2D_3E_1$	HNT level: 0 wt%, nozzle temperature: 220 °C, print speed 20 mm/s, infill density: 100%, layer height: 0.2 mm	4.783
Thickness error (%)	Cylindrical	$A_1B_2C_2D_3E_1$	HNT level: 0 wt%, nozzle temperature: 220 °C, print speed 20 mm/s, infill density: 100%, layer height: 0.2 mm	1.970
Circularity (mm)	Cylindrical	$A_1B_2C_3D_2E_1$	HNT level: 0 wt%, nozzle temperature: 220 °C, print speed 30 mm/s, infill density: 70%, layer height: 0.2 mm	0.164
Cylindricity (mm)	Cylindrical	$A_1B_3C_3D_2E_3$	HNT level: 0 wt%, nozzle temperature: 230 °C, print speed 30 mm/s, infill density: 70%, layer height: 0.4 mm	0.224
Surface roughness (μm)	Cylindrical	$A_1B_1C_2D_3E_1$	HNT level: 0 wt%, nozzle temperature: 210 °C, print speed 20 mm/s, infill density: 100%, layer height: 0.2 mm	4.125

N ratios for a preferred factor-level combination for all DoE responses were used as $S/N_{confirmation}$ listed in Table 8.4. Additionally, calculation results in relation to $S/N_{predicted}$, $CI_{predicted}$ and $CI_{confirmation}$ for dog-bone and cylindrical nanocomposite samples were listed in Table 8.4. As earlier mentioned in Sect. 6.4.4 regarding the confident interval CI column for the evaluation of experimental reproducibility, the intervals of $S/N_{predicted}$ and $CI_{predicted}$ were thus compared with $S/N_{confirmation}$ and $CI_{confirmation}$, which indicated $CI_{predicted}$ and $CI_{confirmation}$ demonstrated a good overlapping level for all DoE responses of dog-bone and cylindrical nanocomposite samples with the confirmation of good test reproducibility in this study.

8.6 Summary

Dog-bone and cylindrical samples based on PU/HNT nanocomposites were optimally fabricated according to ASTM D638 (Type V) and ASTM D5963 standards via FDM. Various HNT contents and 3D printing parameters were selected based on nozzle temperature, print speed, infill density and layer height according to L_{18} OA. DoE responses consisted of gauge section width, thickness, circularity, cylindricity and surface quality of both dog-bone and cylindrical nanocomposite samples. A "smaller-the-better" criterion was followed in this study in response to the lowest dimensional errors and surface roughness of 3D printed nanocomposite samples. HNT content was detected to prevalently influence all DoE response irrespective of dog-bone or cylindrical samples. On the other hand, layer height also played a significant role in thickness, circularity and surface quality of dog-bone and cylindrical nanocomposite samples. As far as HNT content was concerned, print speed induced great impact on the accuracy of gauge section width for dog-bone samples, as well as the accuracy of cylindricity for both sample types. The lowest HNT content of 0 wt% and smallest layer height of 0.2 mm were proven to be beneficial to the best accuracy of thickness, circularity and surface quality of both dog-bone and cylindrical samples. Whereas the lowest error was detected for gauge section width of dog-bone samples at the HNT content of 0 wt% with the medium print speed of 20 mm/s. Finally, the highest accuracy of cylindricity took place with the HNT content of 0 wt% and the highest print speed of 30 mm/s.

Table 8.4 Summary of confirmation tests with respect to *S/N* ratio and confidence interval (*CI*) for nanocomposite dog-bone and cylindrical samples

Dimension	Structure	Prediction		Confirmation		Confidence interval (*CI*)
		$S/N_{predicted}$	$CI_{predicted}$	$S/N_{confirmation}$	$CI_{confirmation}$	
Thickness	Dog-bone	– 13.843	± 0.739	– 13.224	± 1.331	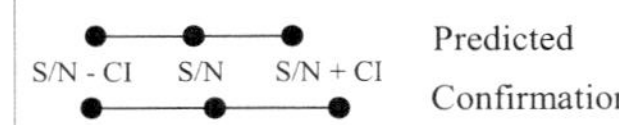
Gauge section width	Dog-bone	– 8.743	± 0.415	– 8.541	± 0.749	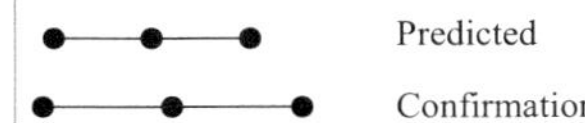
Circularity	Dog-bone	18.909	± 0.352	19.094	± 0.635	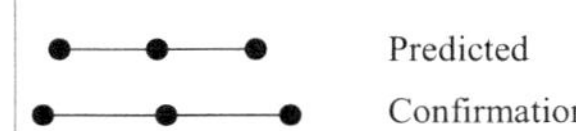
Cylindricity	Dog-bone	14.307	± 0.078	14.379	± 0.140	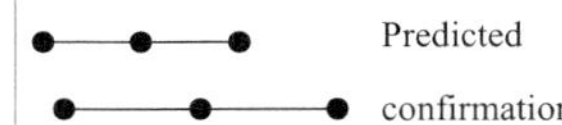
Surface roughness	Dog-bone	– 13.514	± 0.377	– 13.595	± 0.680	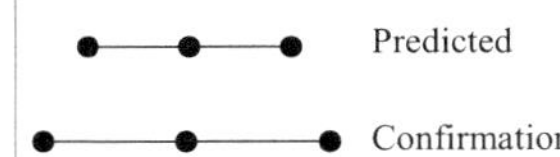
Thickness	Cylindrical	– 5.437	± 1.450	– 5.889	± 2.615	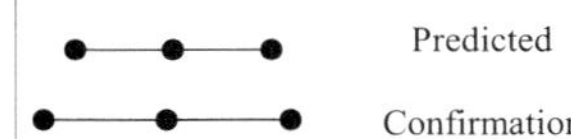
Circularity	Cylindrical	15.350	± 0.386	15.685	± 0.696	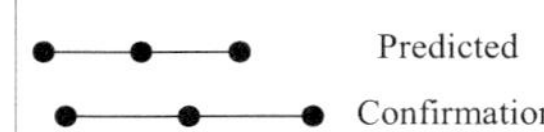

(continued)

Table 8.4 (continued)

Dimension	Structure	Prediction		Confirmation		Confidence interval (CI)
		$S/N_{predicted}$	$CI_{predicted}$	$S/N_{confirmation}$	$CI_{confirmation}$	
Cylindricity	Cylindrical	12.684	± 0.389	12.995	± 0.700	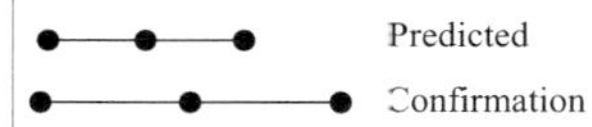
Surface roughness	Cylindrical	– 12.166	± 0.659	– 12.308	± 1.188	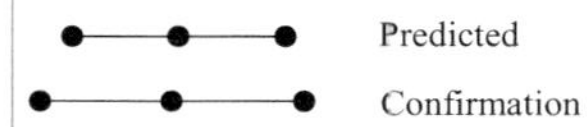

References

1. SMP Technologies: Shape Memory Polymer. SMP Technologies (2021). http://www2.smptechno.com/en/smp/. Accessed 5 Mar 2022
2. Imerys: Halloysite. Imerys (2021). https://www.imerys.com/minerals/halloysite. Accessed 10 Nov 2021
3. Park, S.: Robust Design and Analysis for Quality Engineering. Chapman & Hall, London (1996)
4. Ross, P.: Taguchi Techniques for Quality Engineering, Loss Function, Orthogonal Experiments, Parameter and Tolerance Design. McGraw-Hill, New York (1988)
5. Nugroho, W.T., Dong, Y., Pramanik, A., Zhang, Z., Ramakrishna, S.: Co-influence of nanofiller content and 3D printing parameters on mechanical properties of thermoplastic polyurethane (TPU)/halloysite nanotube (HNT) nanocomposites. Nanomaterials **13**, 1975 (2023)
6. Nugroho, W.T., Dong, Y., Pramanik, A.: Mechanical and shape memory properties of additively manufactured polyurethane (PU)/halloysite nanotube (HNT) nanocomposites. Nanomaterials **14**, 1373 (2024)
7. Poompiew, N., Pattananuwat, P., Aumnate, C., Roman, A., Osswald, T., Potiyaraj, P.: 3D printable resin/carbon nanotube composites for wearable strain sensors: enhancing mechanical and electrical properties. J. Sci. Adv. Mater. Dev. **8**, 100546 (2023)
8. Salkuti, P.: Quality and Performance Evaluation of TPU based Fused Deposition Modeling-Investigations on dimension, surface topography and mechanical properties. Master's thesis, Halmstad University (2018)
9. Zhang, C., Li, P., Chu, F., Shen, G.: Influence of the three-dimensional printing technique and printing layer thickness on model accuracy. J. Orofac. Orthop. **80**, 194–204 (2019)
10. Plaza, E., Lopez, P., Torija, M., Muñoz, J.: Analysis of PLA geometric properties processed by FFF additive manufacturing: Effects of process parameters and plate-extruder precision motion. Polymers **11**, 1581 (2019)
11. Zarko, J., Vladic, G., Pal, M., Dedijer, S.: Influence of printing speed on production of embossing tools using FDM 3D printing technology. J. Graph. Eng. Des. **8**, 19–27 (2017)
12. Pandey, P., Nallagundla, V., Dhande, S.: Real time adaptive slicing for fused deposition modelling. Int. J. Mach. Tools Manuf **43**, 61–71 (2003)

Chapter 9
Effect of HNT Content and 3D Printing Parameters on Mechanical and Thermal Properties of PU/HNT Nanocomposites

Abstract This chapter focussed on the influence of both nanofillers and 3D printing parameters on mechanical and thermal properties of 3D printed PU/HNT nanocomposites according to optimum factor-level combinations using Taguchi design of experiments (DoEs) based on L_{18} orthogonal array. The discussion involved tensile properties, hardness and abrasion resistance of 3D printed nanocomposite samples via fused deposition modelling (FDM). Morphological structures, Fourier transform infrared (FTIR) analysis, thermogravimetric analysis (TGA) and differential scanning calorimetry (DSC) were also reported in detail.

Keywords Fused deposition modelling (FDM) · HNT content · 3D print parameters · PU/HNT nanocomposites · Design of experiments (DoEs) · Mechanical and thermal properties

9.1 Mechanical Tests and Material Characterisation

Dog-bone tensile testing samples were prepared according to ASTM D638 (type V) standard. Tensile tests were conducted at room temperature with the load capacity of 10 kN at the crosshead speed of 1 mm/min. Abrasion resistance tests were also performed at room temperature on cylindrical testing samples based on ASTM D5963 standard. Whereas the hardness of both dog-bone and cylindrical samples was examined using ASTM D2240 standard. Surface morphology and HNT dispersion in nanocomposite samples were assessed via scanning electron microscopy (SEM) and transmission electron microscopy (TEM) respectively. FTIR analysis was carried out to investigate chemical structures of PU, HNTs and corresponding nanocomposites at different HNT contents of 2, 4, 6, 8, and 10 wt%. Furthermore, thermal stability of nanocomposite samples was investigated with the aid of TGA along with the determination of glass transition temperature (T_g), crystallisation temperature (T_c), as well as melting temperature (T_m). In particular, the degree of crystallinity (χ_c) of nanocomposite samples was calculated according to Eq. (9.1) as follows [1]:

W. T. Nugroho et al., *Thermoplastic Polyurethane (TPU)/Halloysite Nanotube (HNT) Nanocomposites*, Engineering Materials, https://doi.org/10.1007/978-981-95-0541-8_9

$$\chi_c(\%) = \frac{\Delta H_{m-}\Delta H_c}{\Delta H_m^0} \times \frac{100}{1 - w_f} \quad (9.1)$$

where ΔH_m and ΔH_c denote the heat of fusion and heat of crystallisation of nanocomposite samples respectively. Meanwhile, $\Delta H_m{}^0$ and w_f represent 100% crystalline PU and the weight fraction of HNTs in nanocomposite samples. $\Delta H_m{}^0$ is generally about 140 J/g for PU [2]. Detailed information regarding mechanical tests and material characterisation was introduced in Sect. 6.3.

9.2 Tensile Properties

In general, tensile strength of 3D printed nanocomposite samples was enhanced at the HNT content of 8 wt% with the highest strength level of 56.34 MPa achieved for TN13, which was equivalent to an increase of 24.2% when compared with that of TN1 (i.e. 0 wt% HNTs) at 45.35 MPa shown in Fig. 9.1a. It was worth noting that a good interaction took place between PU matrices and HNTs as the reinforcements in order to improve hydrogen bonding with the presence of carboxyl groups on HNT surfaces [3]. When increasing the HNT content up to 10 wt%, HNT aggregation was manifested to yield a declining trend of tensile strength. In particular, the lowest tensile strength of 26.11 MPa was identified for TN18 with weak filler-matrix interaction.

The incorporation of HNTs as rigid nanofillers improved tensile moduli of nanocomposite samples [3], as evidenced by Fig. 9.1b. The inclusion of 10 wt% HNTs gave rise to the maximum tensile modulus of 4.91 GPa for TN17, as opposed to minimum modulus of 2.07 GPa for TN1 (i.e. 0 wt% HNTs). The incorporation of rigid fillers in composite materials generally yields the enhancement of their stiffness. HNTs could highly restrict chain mobility of PU molecules as rigid fillers. Hence, with increasing material stiffness, its flexibility became detrimental in good accordance with Sulong et al. [4]. As a result, the elongation at break for dog-bone samples decreased when increasing the nanofiller content according to Fig. 9.1c.

Pareto ANOVA was used to determine the factorial contribution with respect to tensile properties of 3D printed dog-bone nanocomposite samples. It has been previously mentioned in Sect. 6.4.4 that significant factors can be identified when the cumulative contribution of input factors exceeds 90% [5]. As evidently seen in Fig. 9.2a, HNT content (factor A) and infill density (factor D) significantly influenced tensile strength at yield for nanocomposite samples with a cumulative contribution of 99%. The presence of HNTs and existing air gaps between 3D printed layers were detrimental to the tensile strength of nanocomposite samples [6]. In contrast, the effect of nozzle temperature (factor B), print speed (factor C) and layer height (factor E) was negligible, which were deemed non-significant factors in this study even though Vidakis et al. [7] mentioned that increasing printing temperature and layer height would induce negative impact on tensile strength instead.

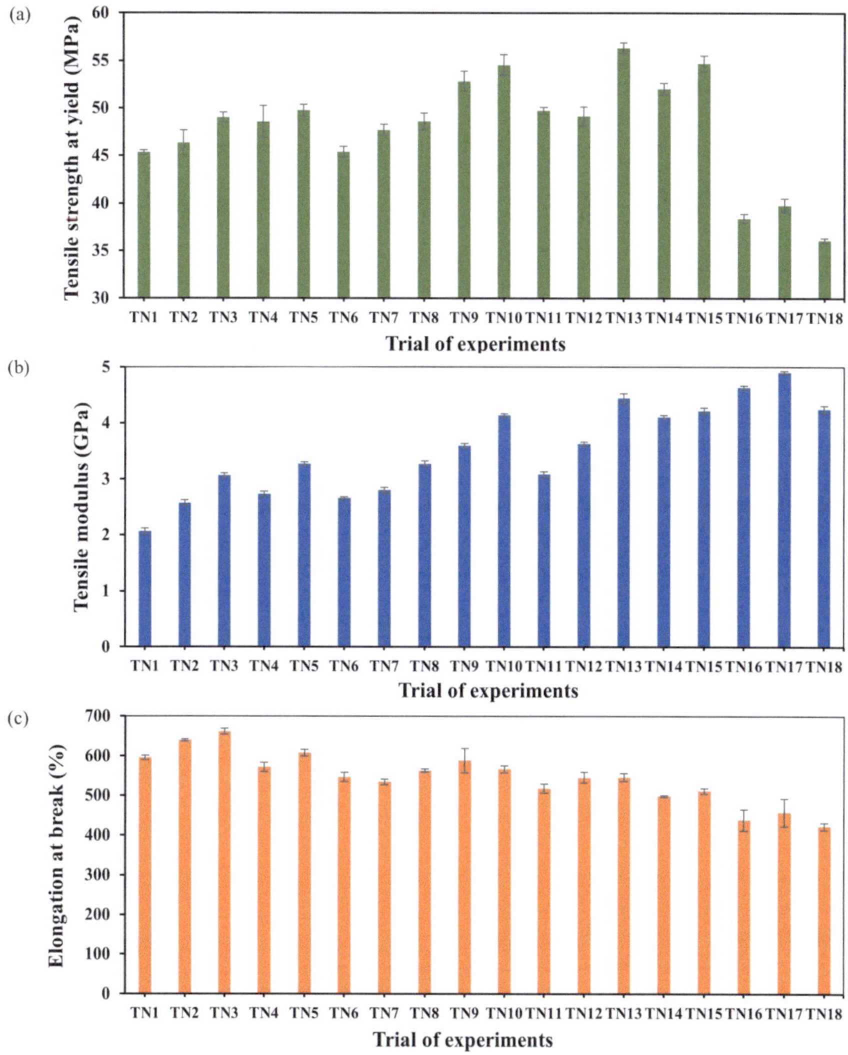

Fig. 9.1 **a** Tensile strength at yield, **b** tensile modulus and **c** elongation at break of 3D printed dog-bone nanocomposite samples using L_{18} DoE

On the other hand, HNT content and infill density also influenced prevalently tensile modulus of 3D printed nanocomposite samples with a cumulative contribution of 96%, as evidenced by Fig. 9.2b. Additionally, such input parameters were also critical with respect to elongation at break based on their corresponding cumulative contribution reaching almost 100% according to Fig. 9.2c. However, other factors including nozzle temperature, print speed and layer height became non-significant,

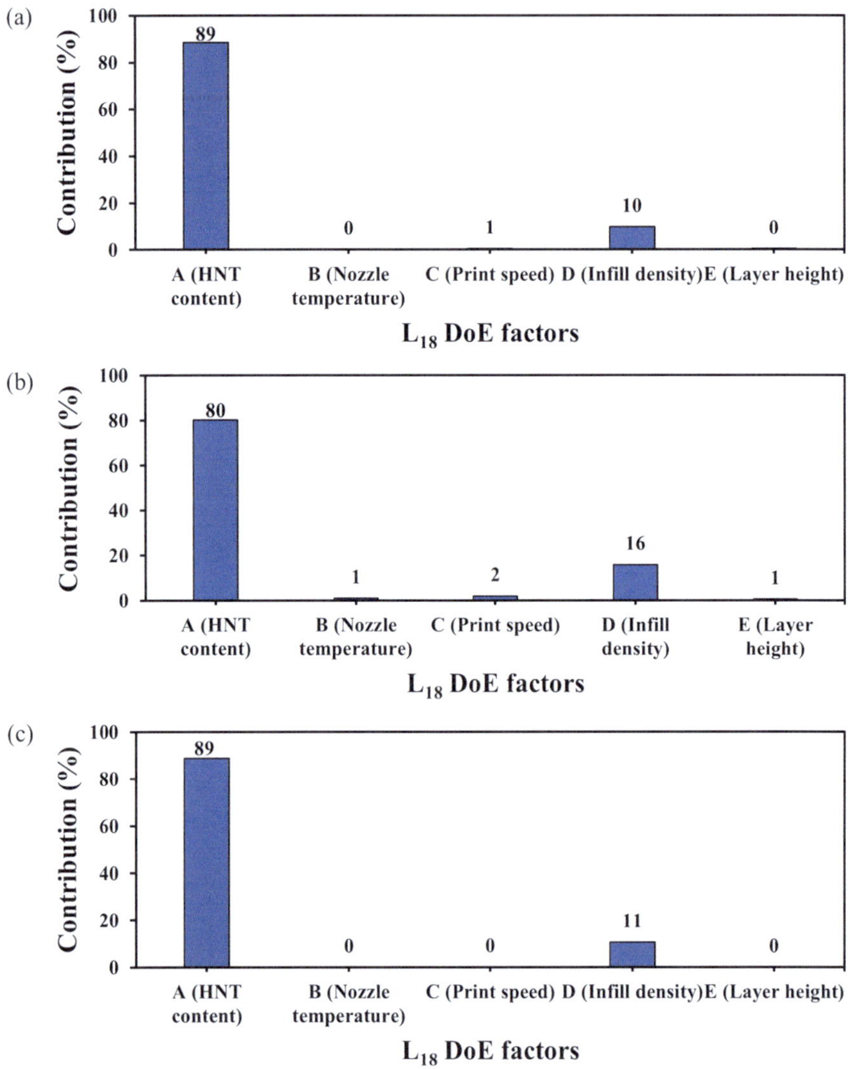

Fig. 9.2 Pareto ANOVA diagrams for enhancing tensile properties of 3D printed dog-bone nanocomposite samples: **a** tensile strength at yield, **b** tensile modulus and **c** elongation at break

which was opposed to Kandi et al. [8] that increasing print speed and layer thickness decreased elongation at break.

In this study, a "larger-the-better" criterion in Taguchi DoE analysis was utilised to assess optimum factor-level combinations in order to maximise mechanical properties of 3D printed nanocomposite samples. Typically, a greater sum of *S/N* ratios signifies a better response to factorial effect. As clearly shown in Fig. 9.3a, increasing

HNT content up to 8 wt% mathematically increased the sum of *S/N* ratios. Nonetheless, when further increasing HNT content to 10 wt%, the sum of *S/N* ratios drastically declined due to HNT aggregation.

Meanwhile, infill density was found to be the second significant factor in response to the maximisation of tensile strength at yield. As expected, the highest tensile

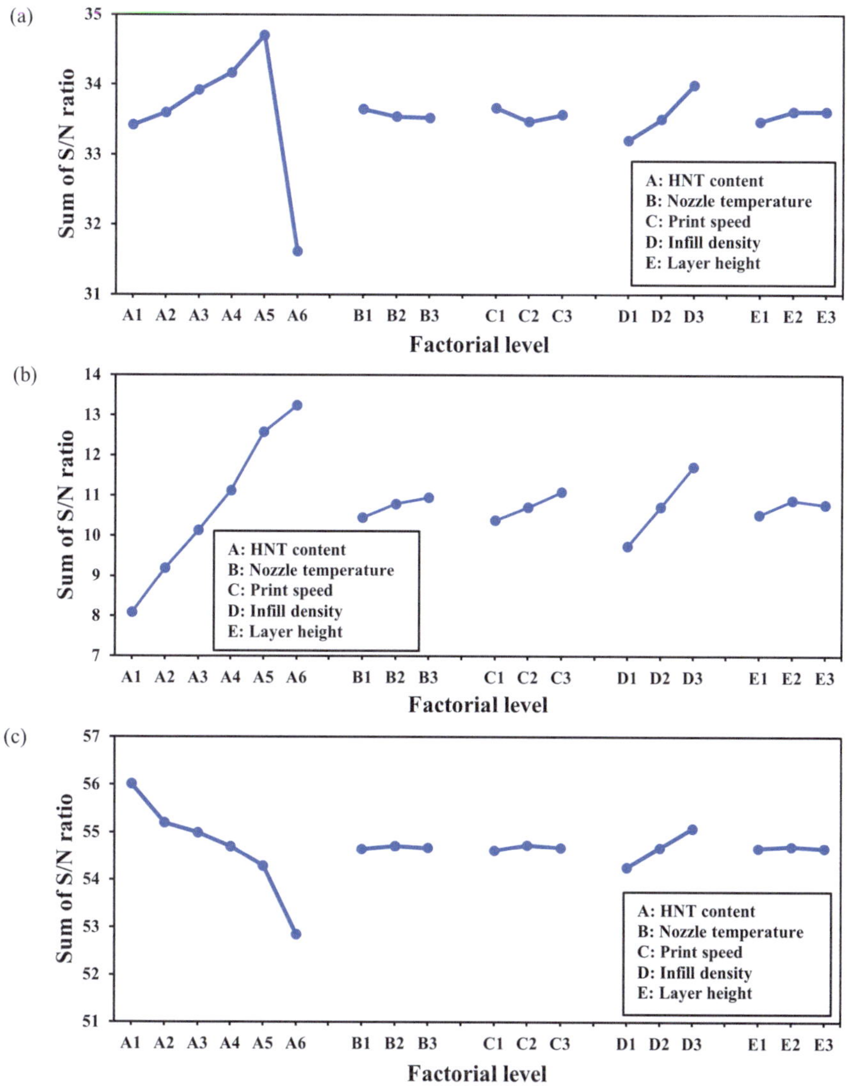

Fig. 9.3 Sum of *S/N* ratio diagrams at different factorial levels with respect to **a** tensile strength at yield, **b** tensile modulus and **c** elongation at break for 3D printed dog-bone nanocomposite samples

strength in this study was achieved at the infill density of 100% for nanocomposite samples in good accordance with Wang et al. [9]. Other non-significant factors consisted of nozzle temperature, print speed and layer height. Ultimately, optimum factor-level combination with respect to maximum tensile strength was obtained with the HNT content of 8 wt%, the nozzle temperature of 210 °C, the print speed of 10 mm/s, the infill density of 100% and the layer height of 0.4 mm (i.e. $A_5B_1C_1D_3E_3$).

Figure 9.3b suggested that the inclusion of HNTs as rigid nanofillers up to 10 wt% within PU matrices increased the stiffness of dog-bone nanocomposite samples, as evidenced by increasing the sum of *S/N* ratios in a monotonic manner. The optimum factor-level combination in response to maximum tensile modulus was referred to as $A_6B_3C_3D_3E_2$ with the HNT content of 10 wt%, the nozzle temperature of 230 °C, the print speed of 30 mm/s, the infill density of 100%, as well as the layer height of 0.3 mm. On the other hand, elongation at break induced an opposite trend mainly associated with HNT content according to Fig. 9.3c. This finding could be ascribed to brittle nature and HNT aggregation in PU/HNT nanocomposites. Le et al. [10] reported that maximum elongation at break was detected at the lowest infill density. Overall, optimum factor-level combination in response to the highest elongation at break comprised the HNT content of 0 wt%, the nozzle temperature of 230 °C, the print speed of 20 mm/s, the infill density of 100%, as well as the layer height of 0.3 mm (i.e. $A_1B_3C_2D_3E_2$).

9.3 Hardness and Abrasion Resistance

The incorporation of 10 wt% HNTs improved Shore D hardness of dog-bone nanocomposite samples from 71.50 for TN1 to 76.67 for TN17, as illustrated in Fig. 9.4a. This result could be attributed to the occurrence of new hydrogen bonds between HNTs and PU molecular chains to prevent these bonds between polymeric segments [3]. Apparently, inherent stiffness and hardness of HNTs contributed to the increase in hardness of nanocomposite samples with the addition of 2 wt% HNTs to reinforce PU matrices in nanocomposites for the purpose of corrosive coating [11].

As seen in Fig. 9.4b, Shore D hardness results of cylindrical samples also demonstrated an increasing tendency with the inclusion of HNTs in PU/HNT nanocomposites. It was determined that minimum and maximum hardness values of cylindrical samples were achieved at approximately 70.61 for TN1 (neat PU), as well as 77.28 for TN17 with 10 wt% HNTs. Corresponding abrasion resistance of cylindrical samples was calculated in terms of volume loss [12]. It was revealed in Fig. 10.4c that greater volume loss and a further decrease in abrasion resistance arose from higher HNT content levels in cylindrical samples. This finding coincided with Mohamed et al. [11] indicating the increase of volume loss to 96% with the inclusion of 2 wt% HNTs when compared with neat PU.

HNT content and infill density were determined to be two significant factors with respect to the hardness of dog-bone nanocomposite samples with their corresponding contributions of 50% and 45% respectively, as illustrated in Fig. 9.5a.

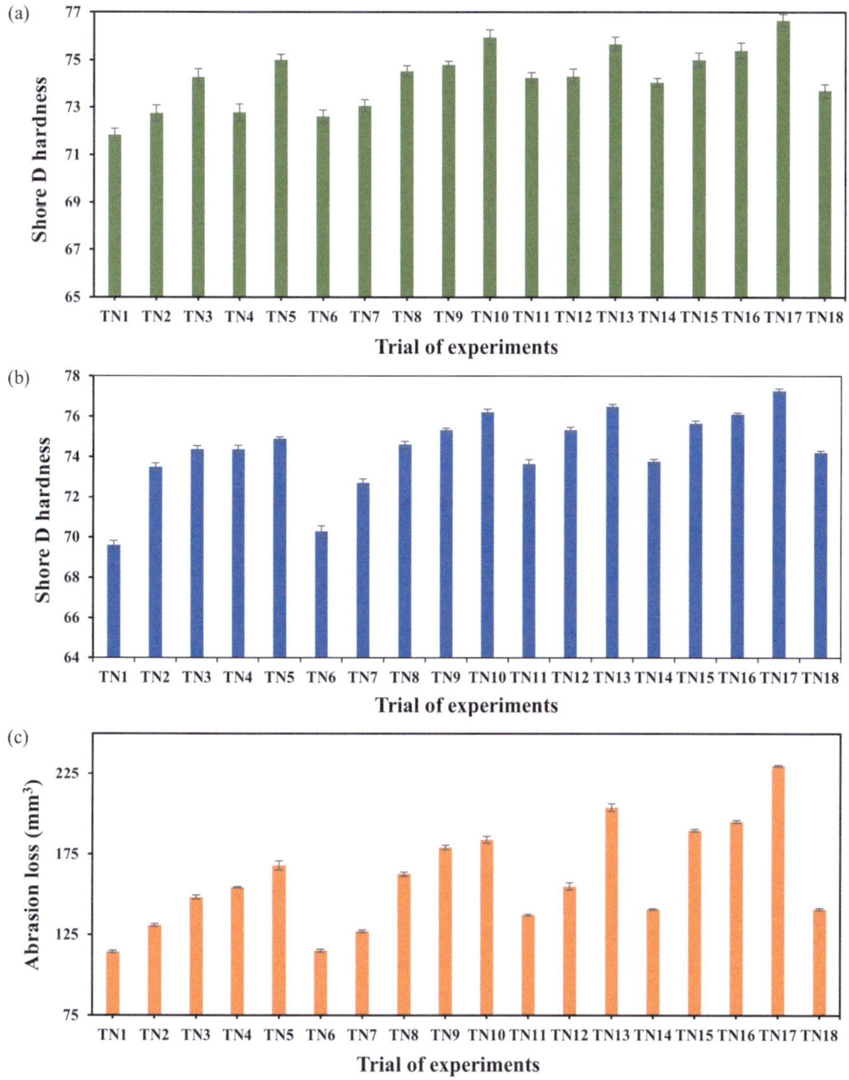

Fig. 9.4 **a** Hardness of dog-bone nanocomposite samples, as well as **b** hardness and **c** abrasion loss of cylindrical nanocomposite samples

Nozzle temperature, print speed and layer height only contributed to 1%, 1% and 2% as non-significant factors accordingly. On the contrary, the hardness of cylindrical nanocomposite samples was significantly influenced by infill density and HNT content with a cumulative contribution of 93% depicted in Fig. 9.5b. In comparison, nozzle temperature, print speed and layer height only contributed to 2%, 2% and 3% respectively. As shown in Fig. 9.5c, main contributions of 59% and 37% arose from infill density and HNT content with trivial effect from other input factors. In this study using Taguchi DoEs, a "larger-the-better" criterion was followed to determine optimum factor-level combination in response to the greatest hardness of dog-bone and cylindrical nanocomposite samples.

Increasing infill density, as evidenced by Fig. 9.6a, increased the sum of *S/N* ratios for the hardness of dog-bone samples due to fewer air gaps identified between deposited layers, thus resulting in much denser materials with higher deformation

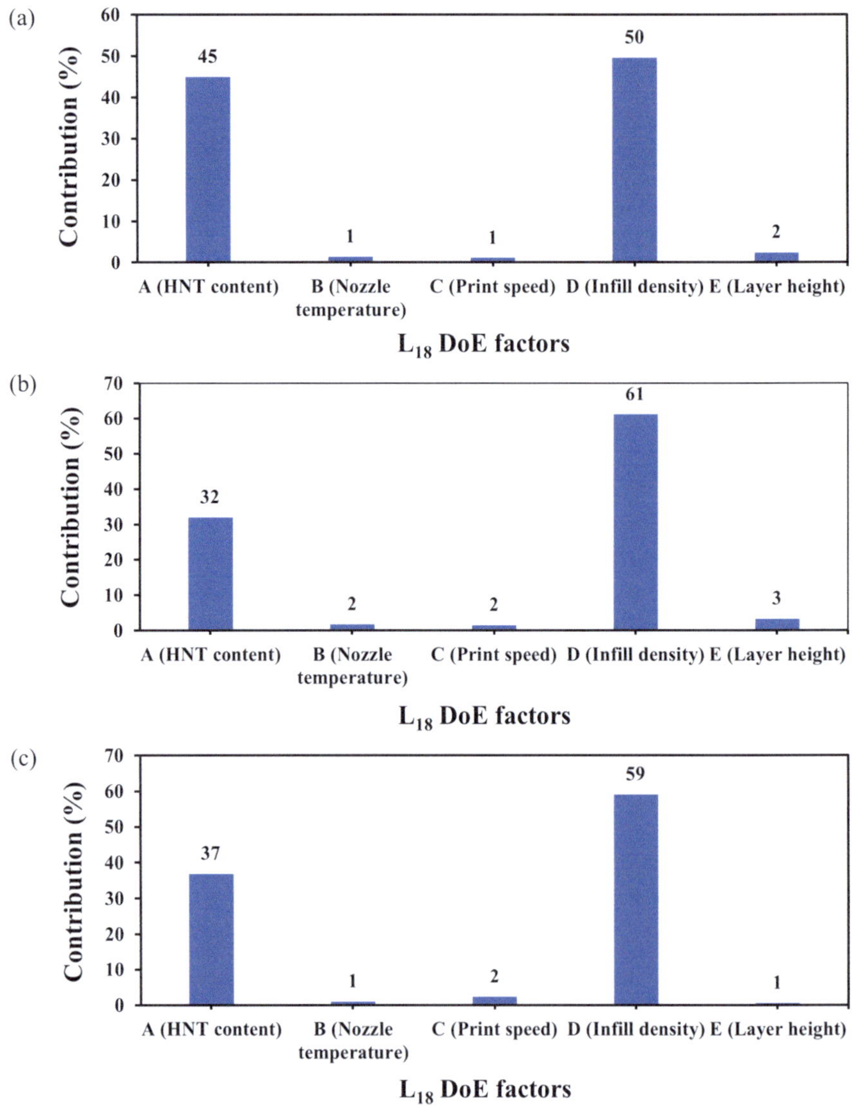

Fig. 9.5 Pareto ANOVA diagrams for maximising **a** hardness of dog-bone nanocomposite samples, as well as **b** hardness and **c** abrasion resistance of cylindrical nanocomposite samples

resistance [13]. The inclusion of HNTs was believed to yield more compacted and stronger hydrogen bonds between HNTs and PU molecular chains. Optimum factor-level combination was referred to as $A_6B_2C_3D_3E_3$ with the HNT content of 10 wt%, the nozzle temperature of 220 °C, the print speed of 30 mm/s, the infill density of 100%, as well as the layer thickness of 0.4 mm.

As illustrated in Fig. 9.6b, optimum factor-level combination was comprised of the HNT content of 10 wt%, the nozzle temperature of 220 °C, the print speed of 20 mm/s, as well as the layer height of 0.4 mm (i.e. $A_6B_2C_2D_3E_3$). As earlier mentioned, abrasion resistance depends primarily on material resistance to wear deformation where better abrasion durability should be characterised by smaller abrasion loss [14]. As such, a "smaller-the-better" criterion was followed in DoE work to determine the optimum factor-level combination in response to the lowest abrasion loss of cylindrical nanocomposite samples. Lower infill density and HNT content resulted in higher abrasion resistance, which signified a completely opposite trend to hardness. Overall, the highest abrasion resistance was obtained with the HNT content of 0 wt%, the nozzle temperature of 230 °C, the print speed of 20 mm/s, the infill density of 40%, as well as the layer height of 0.3 mm (i.e. $A_1B_3C_2D_1E_2$).

Apparently, the incorporation of HNTs as rigid nanofillers within PU matrices enhanced the hardness of PU/HNT nanocomposites. It was shown in Fig. 9.7 that abrasion loss was enhanced monotonically with increasing the hardness of nanocomposite samples, as evidenced by *R*-square (R^2) at 83% using a linear regression. On the contrary, Beck et al. [15] reported that materials with higher hardness tended to be more resilient to abrasive wear in good agreement with Archard's wear law [16]. However, it was contradictory to several other studies associated with abrasion behaviour [17–20] that abrasive wear resistance of PU was promoted when increasing the content of soft segments. Additionally, the quantity of soft segments was reduced with increasing nanofiller content, leading to an increase in abrasion loss, which meant the deterioration of abrasive wear resistance in this study. Complete optimum factor-level combinations to improve mechanical properties of nanocomposites samples for dog-bone and cylindrical nanocomposite samples were presented in Table 9.1, along with confirmation test results. The overall optimum factor-level combinations to improve mechanical properties of nanocomposite samples, as well as the confirmation test results for dog-bone and cylindrical samples were presented in Table 9.1.

As for mechanical properties of nanocomposite samples, tensile strength is critical to achieve desired material performance and efficient load-bearing capacity of a material. As such, optimum factor-level combination in response to maximum tensile strength at yield was chosen according to Table 9.2. Whereas other printing factors were fixed in Table 9.3 to fabricate nanocomposite samples based on PU (MM4520 grade) pellets and HNTs. These samples were further manufactured shown in Tables 9.2 and 9.3 at different HNT contents ranging from 0, 2, 4, 6, 8 to 10 wt% in order to assess their tensile and thermal properties, as well as chemical and morphological structures.

Typical stress–strain curves associated with mechanical properties of nanocomposite samples in terms of tensile strength, tensile modulus and elongation at break

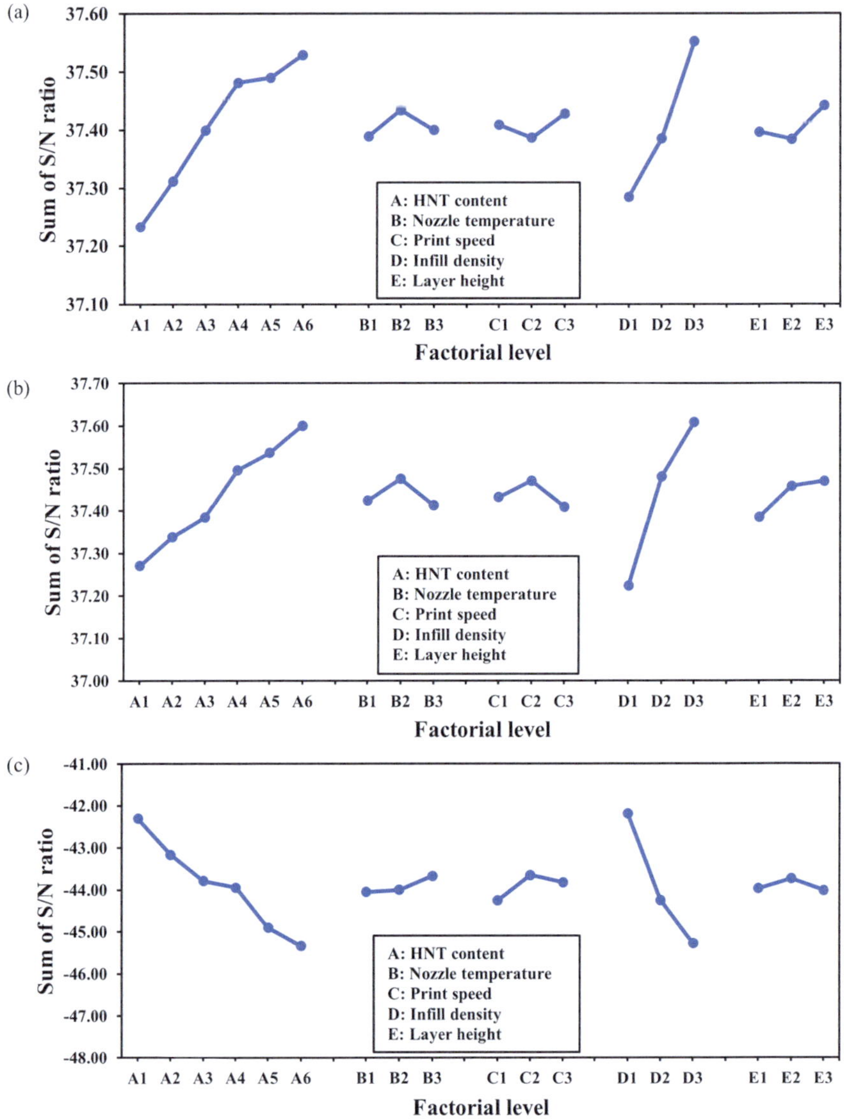

Fig. 9.6 Sum of *S/N* ratio diagrams at different factorial levels for **a** hardness of dog-bone nanocomposite samples, as well as **b** hardness and **c** abrasion loss of cylindrical nanocomposite samples

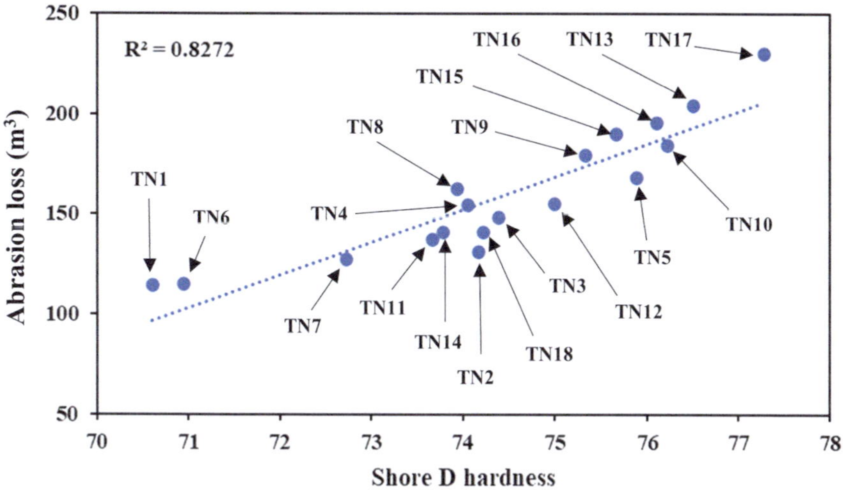

Fig. 9.7 A linear relationship between hardness and abrasion loss of cylindrical nanocomposite samples

were depicted in Figs. 9.8 and 9.9 respectively. Overall, the incorporation of 8 wt% HNTs in nanocomposite samples increased tensile strength of nanocomposite samples by 30.4% (up to 58.34 MPa), as opposed to 44.75 MPa for neat PU according to Fig. 9.9a. Such an increase could be associated with better filler-matrix interfacial bonding with the presence of surface functional groups [3]. Consequently, more energy was required to stretch nanocomposite samples than neat PU. However, when increasing HNT content up to 10 wt%, the interaction between PU matrices and HNTs became detrimental, as evidenced by a decrease in tensile strength to 40.61 MPa due to typical HNT aggregation.

Furthermore, the inclusion of HNTs as rigid nanofillers within PU matrices enhanced tensile modulus of nanocomposite samples according to Fig. 9.9b. The highest tensile modulus of 3.37 GPa was achieved at the HNT content of 10 wt% with a modulus increase up to approximately 47.2% when compared with the minimum modulus of 2.29 GPa for neat PU. This finding may be attributed to stiffening effect of rigid HNTs to restrict chain mobility of PU molecules. As a result, it increased the stiffness of nanocomposites, which was followed by decreasing extensibility and flexibility of PU as expected [4]. In contrast, the flexibility appeared to reduce elongation at break for nanocomposite samples with increasing HNT content. It was observed in Fig. 9.9c that a sudden decline occurred when HNT content exceeded 2 wt%. This result was expected since the presence of rigid nanofillers such as HNTs inevitably enhanced the brittleness of polymer nanocomposites. It resulted in the minimum elongation at break for nanocomposites at the HNT content of 10 wt%, which was equivalent to a 28.7% decrease relative to that of neat PU. Tensile properties of nanocomposite samples in this study were in good accordance with previous results reported by Namathoti et al. [21].

Table 9.1 Summary of optimum factor-level combinations for enhanced mechanical properties of nanocomposite samples

Larger-the-better L_{18} DoE response	Significant factor	Optimum combination	Factor-level combination	Confirmation test
Tensile strength at yield (MPa)	• HNT content • Infill density	$A_5B_1C_1D_3E_3$	• HNT content (wt%): 8 • Nozzle temperature (°C): 210 • Print speed (mm/s): 10 • Infill density (%): 100 • Layer height (mm): 0.4	57.293
Tensile modulus (GPa)	• HNT content • Infill density	$A_6B_3C_3D_3E_2$	• HNT content (wt%): 10 • Nozzle temperature (°C): 230 • Print speed (mm/s): 30 • Infill density (%): 100 • Layer height (mm): 0.3	5.003
Elongation at break (%)	• HNT content • Infill density	$A_1B_3C_2D_3E_2$	• HNT content (wt%): 0 • Nozzle temperature (°C): 230 • Print speed (mm/s): 20 • Infill density (%): 100 • Layer height (mm): 0.3	668.68
Shore D hardness of dog-bone nanocomposite samples	• Infill density • HNT content	$A_6B_2C_3D_3E_3$	• HNT content (wt%): 10 • Nozzle temperature (°C): 220 • Print speed (mm/s): 30 • Infill density (%): 100 • Layer height (mm): 0.4	76.867

(continued)

Table 9.1 (continued)

Larger-the-better L_{18} DoE response	Significant factor	Optimum combination	Factor-level combination	Confirmation test
Shore D hardness of the cylindrical nanocomposite sample	• Infill density • HNT content	$A_6B_2C_2D_3E_3$	• HNT content (wt%): 10 • Nozzle temperature (°C): 220 • Print speed (mm/s): 20 • Infill density (%): 100 • Layer height (mm): 0.4	77.880
Smaller-the-better L_{18} DoE response	**Significant factor**	**Optimum combination**	**Factor-level combination**	**Confirmation test**
Abrasion loss (mm^3)	• Infill density • HNT content	$A_1B_3C_2D_1E_2$	• HNT content (wt%): 0 • Nozzle temperature (°C): 230 • Print speed (mm/s): 20 • Infill density (%): 40 • Layer height (mm): 0.3	112.813

Table 9.2 Preferred 3D printing parameters

Parameter	Setting
Nozzle temperature (°C)	210
Print speed (mm/s)	10
Infill density (%)	100
Layer height (mm)	0.4

9.4 Morphological Structures and Nanofiller Dispersion

SEM and energy dispersive X-ray spectroscopic (EDS) analyses were implemented to investigate the morphology and organic components of both PU and PU/HNT nanocomposites according to Fig. 9.10. As shown in Fig. 9.10a, three identified elements, comprising C, N and O, were attributed to organic components of PU. Meanwhile, other elements such as Al and Si could be associated with embedded HNTs with their chemical composition of $Al_2Si_2O_5(OH)_4 \cdot nH_2O$ in nanocomposite samples presented in Fig. 9.10b–f. Apparently, HNTs were well distributed within PU matrices, which was particularly the case for low HNT contents ranging from 2, 4 to 6 wt% according to Figs. 9.10b–d. However, some localised HNT aggregates

Table 9.3 Fixed 3D printing parameters for nanocomposite samples

Parameter	Specific parameter	Setting
Quality	Shell thickness (mm)	1.6
	Initial layer thickness (mm)	0.5
	Initial layer line width (%)	120
	Top surface quality	Precise
Fill	Bottom/top thickness (mm)	1.2
	Infill interface density	Dense
	Infill type	triangle
	Infill overlap (%)	15
Temperature	Bed temperature (°C)	55
Speed	Travel speed (mm/s)	150
	Bottom layer speed (mm/s)	15
	Infill speed (mm/s)	30
Filament	Flow (%)	115
Retraction	Speed (mm/s)	30
	Distance (mm)	5
	Minimum travel (mm)	1.5
	Minimal extrusion before retracting (mm)	0.005

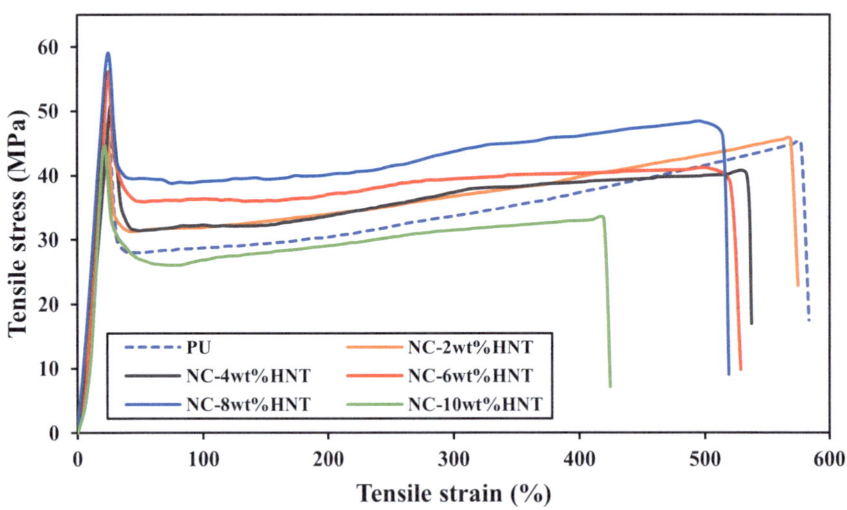

Fig. 9.8 Typical stress–strain diagram of PU/HNT nanocomposites at different HNT contents

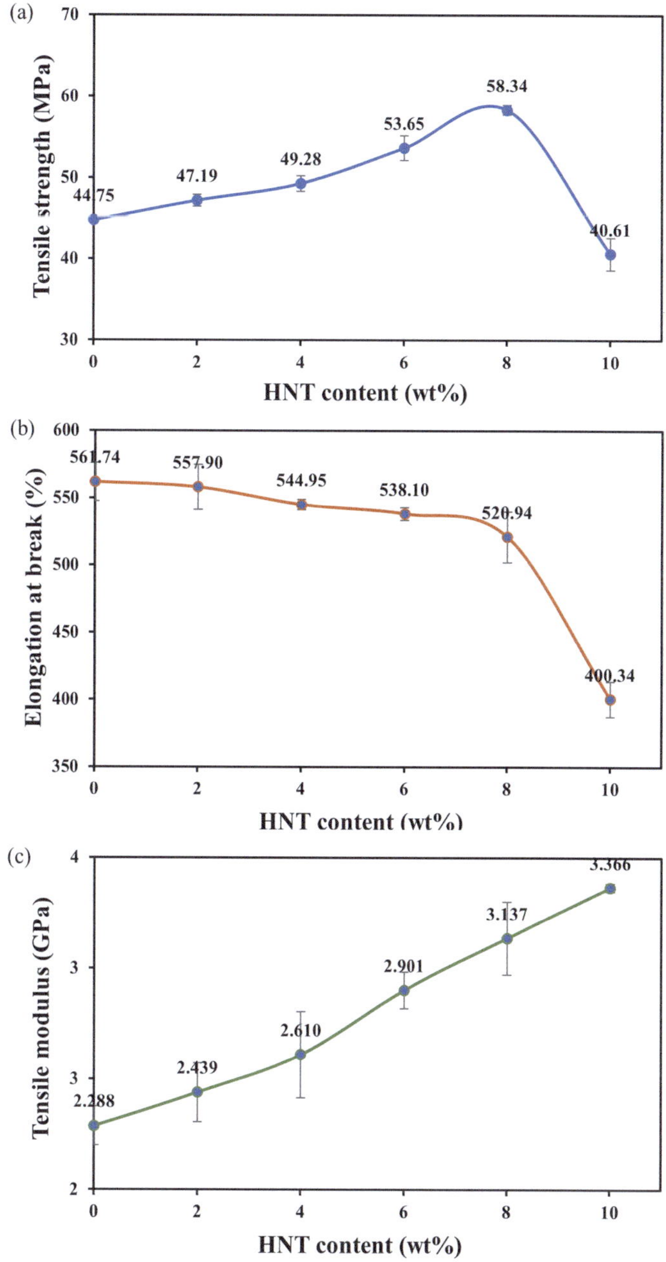

Fig. 9.9 Effect of HNT content on **a** tensile strength, **b** elongation at break and **c** tensile modulus of PU/HNT nanocomposite samples

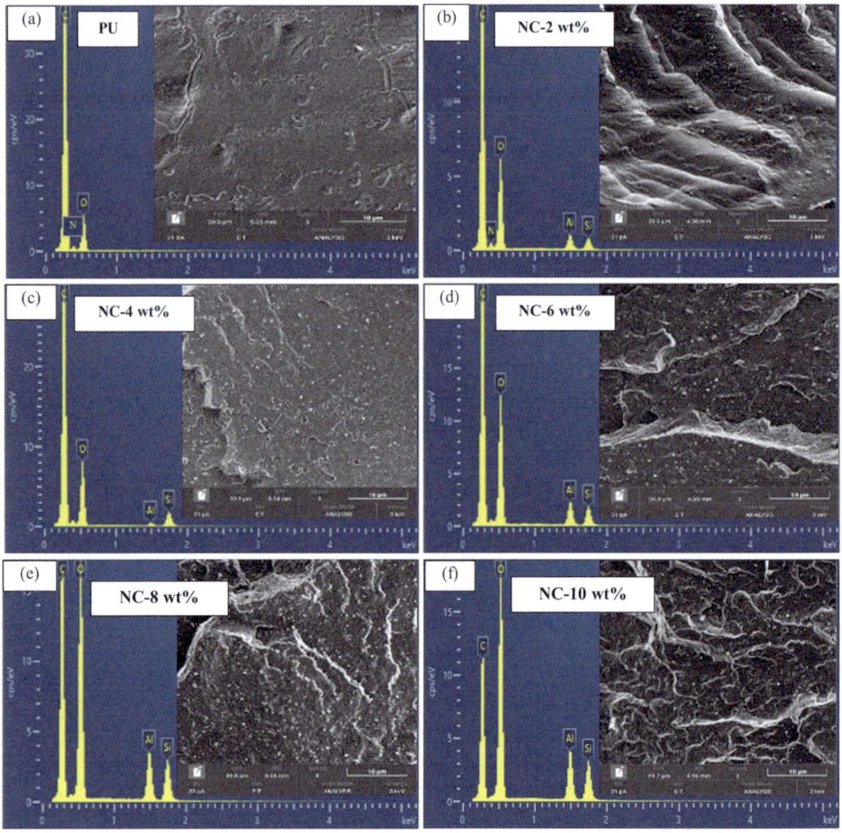

Fig. 9.10 EDS spectra and SEM micrographs of PU/HNT nanocomposites at different contents: **a** 0 wt% (neat PU), **b** 2 wt%, **c** 4 wt%, **d** 6 wt%, **e** 8 wt% and **f** 10 wt%

took place at higher HNT contents of 8 and 10 wt% shown in Fig. 9.10e and f, which were also observed by TEM micrographs depicted in Fig. 9.11a–e.

9.5 FTIR Analysis

FTIR spectra of HNTs, PU and PU/HNT nanocomposites at different HNT contents were illustrated in Fig. 9.12. Nanocomposite samples reinforced with 2, 4, 6, 8 and 10 wt% were referred to as NC-2 wt% HNT, NC-4 wt% HNT, NC-6 wt% HNT, NC-8 wt% HNT and NC-10 wt% HNT respectively. FTIR peaks related to 3690 and 3620/cm were assigned to HNTs for stretching vibration detected on the inner surfaces of hydroxyl groups (Al–OH), as mentioned by Mohammadzadeh et al. [22]. Meanwhile,

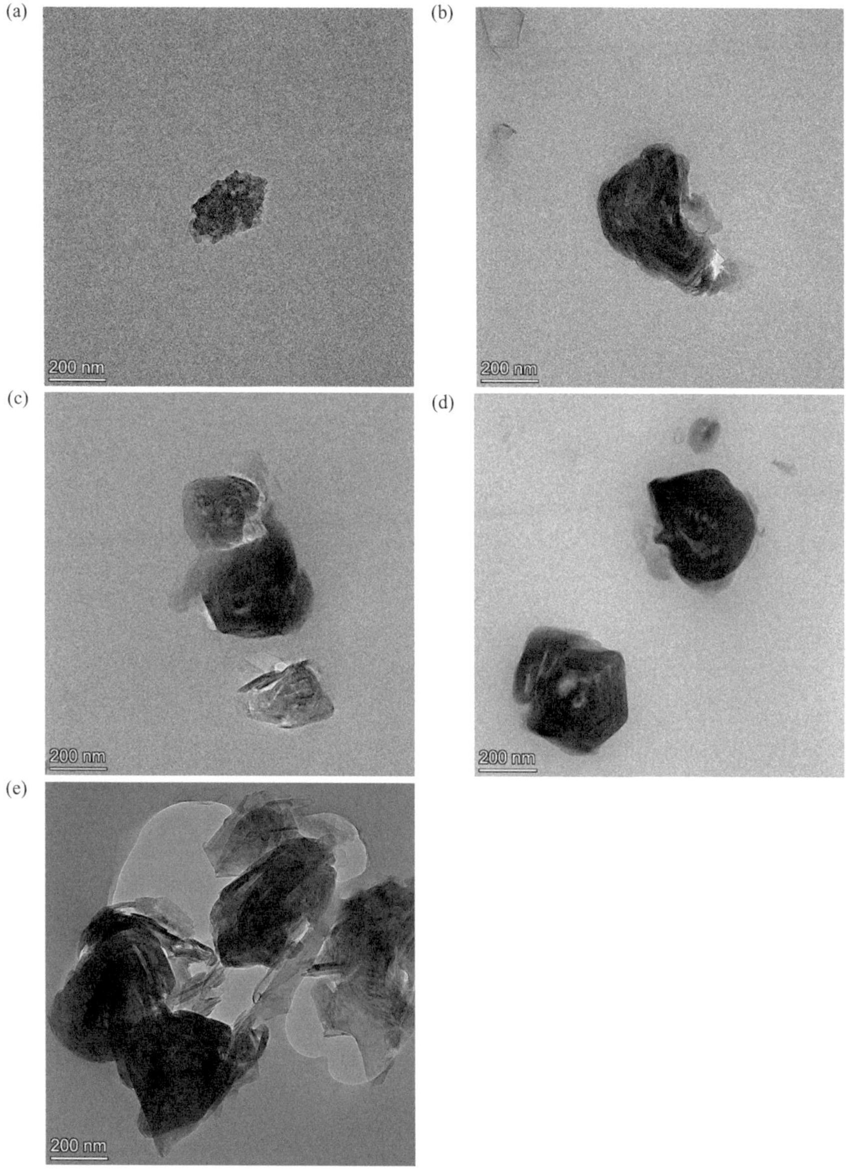

Fig. 9.11 TEM micrographs indicating HNT dispersion within PU matrices in nanocomposites at HNT contents of **a** 2, **b** 4, **c** 6, **d** 8 and **e** 10 wt%

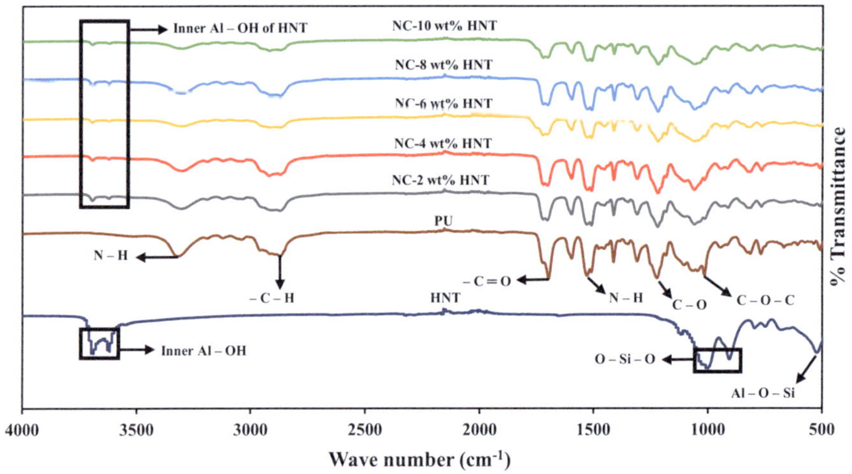

Fig. 9.12 FTIR spectra of HNT, PU and PU/HNT nanocomposites at different HNT contents

FTIR peaks observed at 999 and 905/cm arose from the stretching vibration of O–Si–O [23]. Furthermore, another peak at 523/cm might be due to deformation vibration of Al–O–Si [24].

FTIR peak located at 3317/cm was associated with N–H bond stretching of PU [25]. On the other hand, the absorption band detected at 2869/cm may be related to –CH_2 stretching [26]. The peak identified at 1697/cm was indicative of hydrogen bonds between N–H and C=O groups in hard segments and ester groups in soft segments for urethane linkage [27]. The other three FTIR peaks were attributed to N–H bending vibration, C–O stretching vibration and ester C–O–C symmetric stretching vibration [25]. FTIR spectra for all nanocomposite samples were very similar, which implied almost no direct influence on the alteration of chemical structures regardless of different HNT contents in PU/HNT nanocomposites.

9.6 Thermal Stability

Thermal stability of PU/HNT nanocomposite samples was indicated by their TGA thermograms illustrated in Fig. 9.13a. There were four decomposition temperatures, namely T_{d1}, T_{d2}, T_{d3} and T_{d4}, which were related to the degradation of hard segments within PU matrices into alcohol and isocyanate with a clear sign of process complexity [28]. T_{d1} and T_{d2} were related not only to the dissociation of hard segments of PU/HNT nanocomposites [25], but also to a bond release of water molecules in empty tubular lumens of HNTs [3]. As for T_{d3} and T_{d4}, they were associated with the thermal decomposition of PU soft segments [29] and carbon degradation.

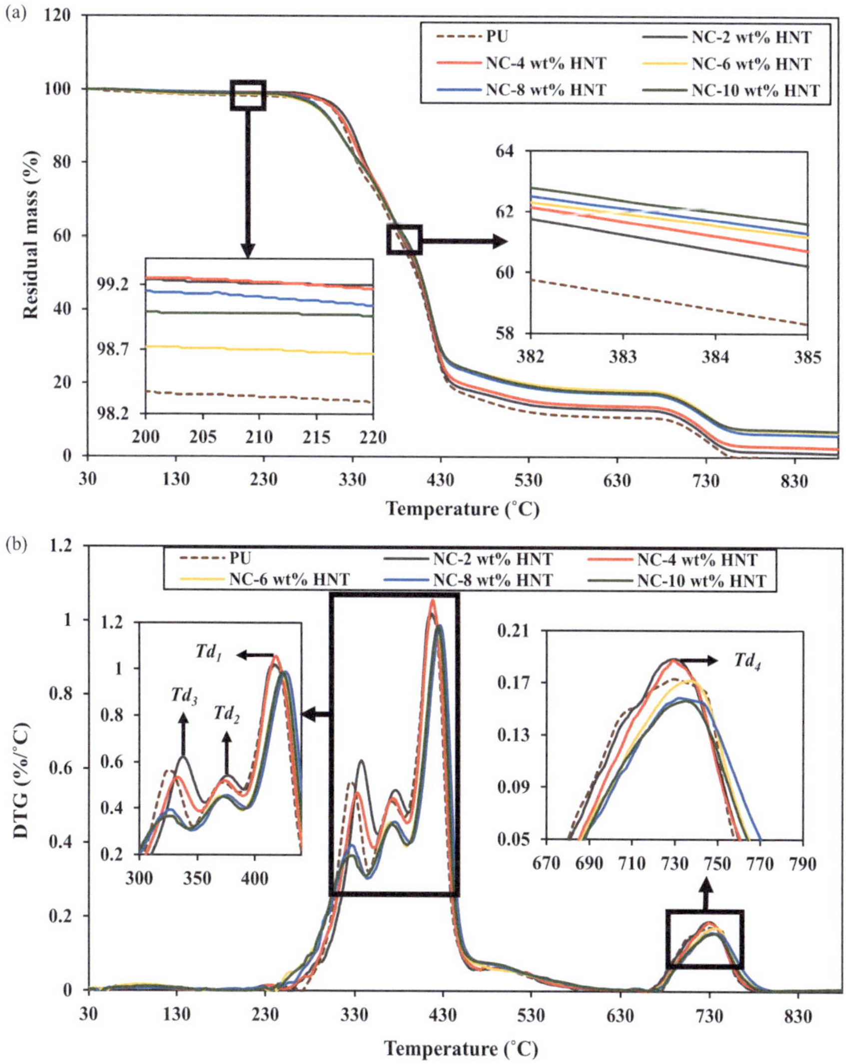

Fig. 9.13 TGA thermograms for PU and PU/HNT nanocomposite samples: **a** TGA curves and **b** DTG curves. T_d refers to the decomposition temperature

Overall, there were four TGA peaks with respect to neat PU at the temperature ranges of 41–348 °C, 348–387 °C, 387–460 °C and 663–788 °C respectively. In contrast, four temperature ranges also appeared at 28–358 °C, 358–391 °C, 391–465 °C and 664–782 °C in relation to four corresponding TGA peaks of PU/HNT nanocomposites at the HNT content of 2 wt%. The inclusion of HNTs within PU matrices shifted T_{d1}, T_{d2}, T_{d3} and T_{d4} to higher temperature levels when compared with those of neat PU. This phenomenon may be induced by barrier effect of HNTs

Table 9.4 TGA data summary of PU and PU/HNT nanocomposite samples

Material type	Sample batch	T_{d1} (°C)	DTG (%/°C)	T_{d2} (°C)	DTG (%/°C)	T_{d3} (°C)	DTG (%/°C)	T_{d4} (°C)	DTG (%/°C)
Neat PU	PU	325.31	0.5661	371.19	0.5118	416.85	1.021	730.08	0.1732
2 wt% HNTs	NC-2 wt% HNT	337.59	0.6223	376.4	0.5418	417.49	1.017	730.15	0.1882
4 wt% HNTs	NC-4 wt% HNT	333.04	0.5336	373.41	0.5196	418.13	1.056	729.12	0.1873
6 wt% HNTs	NC-6 wt% HNT	327.52	0.3655	372.4	0.4496	424.3	0.9843	734.81	0.1563
8 wt% HNTs	NC-8 wt% HNT	326.49	0.393	375.97	0.4568	426.36	0.9878	730.2	0.163
10 wt% HNTs	NC-10 wt% HNT	325.82	0.3656	370.47	0.455	423.82	0.9619	737.3	0.1718

to delay the dissociation of PU hard segments, thus leading to increasing T_d [29]. Furthermore, an entrapping mechanism of HNT lumens for degradation products of polymers enhances thermal stability and major delay in mass transport [30]. Once HNT content exceeded 2 wt%, the decomposition temperatures of nanocomposite samples slightly dropped, especially for T_{d3} and T_{d4} of PU/HNT nanocomposites at HNT contents of 4, 6, 8 and 10 wt% according to Table 9.4. It could be associated with inhibited hydrogen bonding between hard segment domains of PU matrices and embedded HNTs to prevent the retention of thermal stability for polymer nanocomposites [25]. Overall, T_{d1}, T_{d2}, T_{d3} and T_{d4} of nanocomposite samples became higher than those of neat PU, which signified that the incorporation of HNTs tended to enhance thermal stability of resulting PU/HNT nanocomposites.

9.7 DSC Thermal Analysis

DSC thermograms of nanocomposite samples at different HNT contents were presented in Fig. 9.14, along with their thermal parameters listed in Table 9.5. As mentioned earlier, PU molecules consist of both hard and soft segments, while T_g associated with hard segments is due to covalent bonding of NCO and OH groups, enabling to affect the movement of polymeric chains [31]. The addition of HNTs apparently promotes covalent bonding despite the restriction to the chain movement of PU molecules leading to increasing T_g [22]. The presence of 4 wt% HNTs within PU matrices increased T_g up to 42.56 °C, as opposed to 40.96 °C for neat PU shown in Fig. 9.14a and Table 9.5. Beyond this HNT content level, urethane stability was disrupted with a further decrease in T_g [25]. The incorporation of HNTs within PU matrices caused microphase separation owing to typical transformation into hard segments by embedded HNTs [22]. The addition of small amounts of HNTs as much as 2 wt% in nanocomposites yielded the increase of melting temperature up to 166.16 °C when compared with that of neat PU at 161.32 °C. Beyond this content level, the purification of soft segments prevailed along with the extraction of hard segments from the mixture. As such, T_m shifted to the lower temperature level according to Table 9.5, which was in good agreement with Razzaq et al. [32].

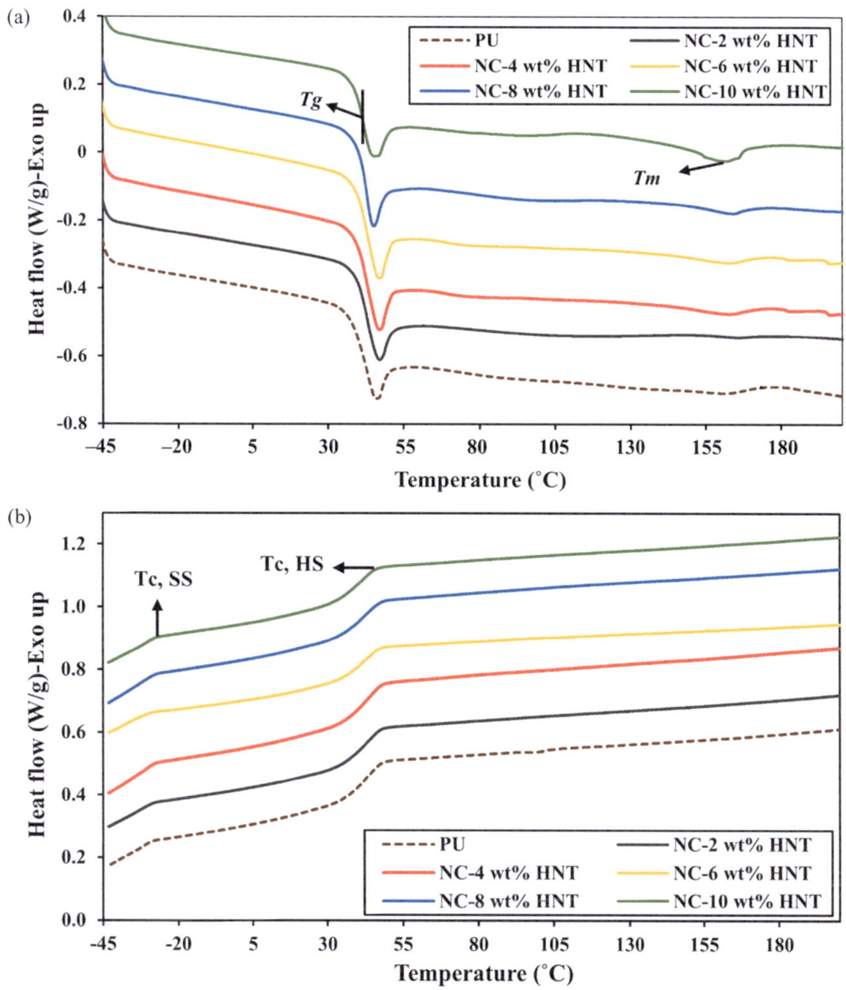

Fig. 9.14 DSC thermograms of nanocomposite samples at different HNT contents: **a** heating and **b** cooling processes

Table 9.5 DSC thermal properties of neat PU and PU/HNT nanocomposites

Material type	Sample batch	T_g (°C)	$T_{c,SS}$ (°C)	$T_{c,HS}$ (°C)	T_m (°C)	ΔH_m (J/g)	ΔH_c (J/g)	χ_c (%)
Neat PU	PU	40.96	– 31.05	47.66	161.32	4.34	0.37	2.84
2 wt% HNT	NC-2 wt% HNT	42.44	– 30.12	48.44	166.16	4.26	0.34	2.85
4 wt% HNT	NC-4 wt% HNT	42.56	– 29.88	49.36	163.32	5.18	0.64	3.38
6 wt% HNT	NC-6 wt% HNT	42.35	– 28.56	49.43	163.04	6.23	0.43	4.41
8 wt% HNT	NC-8 wt% HNT	41.57	– 28.07	48.42	164.03	6.43	0.49	4.61
10 wt% HNT	NC-10 wt% HNT	41.28	– 27.24	48.04	161.78	6.68	0.53	4.89

The addition of HNTs in nanocomposites gave rise to a higher degree of crystallinity χ_c due to typical nucleating effect of HNTs [22]. Once HNT content increased up to 6 wt%, χ_c of nanocomposite samples was enhanced by 55% relative to that of neat PU according to Table 9.5. A further increase in HNT content from 6 to 10 wt% yielded the reduction in the growth rate of nuclei in the crystallisation process, as evidenced by approximately 10% increase in χ_c. On the other hand, the presence of HNTs also shifted DSC peaks with respect to the crystallisation temperatures of hard and soft segments, which were referred to as $T_{c,HS}$ and $T_{c,SS}$ respectively, as shown in Fig. 9.14b and Table 9.5. A higher T_c was proven to enhance microphase separation, thus resulting in pure soft segments and better shape fixity [22, 33].

9.8 Summary

Mechanical and thermal properties of 3D printed nanocomposite samples via FDM with optimum factor-level combinations based on DoEs were successfully investigated in this chapter. The inclusion of 8 wt% HNTs within PU matrices improved both tensile strength and tensile modulus of nanocomposite samples up to 30.4% and 47.2% respectively when compared with those of neat PU. This benefited from better interfacial bonding between surface functional groups of HNTs and hard segments of PU with the incorporation of rigid HNTs. On the contrary, excessive amounts of HNTs up to 10 wt% tended to decrease tensile strength to 40.61 MPa, which might be related to typical HNT agglomeration and weak filler-matrix interaction. Meanwhile, the tensile modulus of nanocomposite samples increased up to 47% when the

HNT content reached 10 wt%. This result was associated with the restricted movement of PU molecular chains because of embedded rigid nanofillers, thus enhancing nanocomposite stiffness along with the reduction of flexibility. As such, PU/HNT nanocomposites appeared to be more brittle, as evidenced by decreasing elongation at break by 400.34% relative to 561.74% for neat PU.

The presence of HNTs also promoted thermal stability of PU/HNT nanocomposites, which was indicative of their relatively high decomposition temperatures. This phenomenon could be attributed to barrier effect of HNTs to delay the degradation of PU hard segments. In addition, an entrapping mechanism of HNT lumens made a significant role in improving thermal stability of nanocomposite samples. Finally, the presence of HNTs also increased T_g, T_m, T_c and χ_c in relation to much better covalent bonding, microphase separation and nucleating effect of HNTs in addition to movement restriction of PU molecules.

References

1. Deng, S., Zhang, J., Ye, L., Wu, J.: Toughening epoxies with halloysite nanotubes. Polymer **49**, 5119–5127 (2008)
2. Cao, H., Qi, F., Liu, R., Wang, F., Zhang, C., Zhang, X., Chai, Y., Zhai, L.: The influence of hydrogen bonding on N-methyldiethanolamine-extended polyurethane solid–solid phase change materials for energy storage. RSC Adv. **17**, 11244–11252 (2017)
3. Mahunaki, M., Asl, V., Mamaqani, H., Koosha, M., Yazdi, M.: Halloysite-reinforced thermoplastic polyurethane nanocomposites: physico-mechanical, rheological, and thermal investigations. Polym. Compos. **41**, 3260–3270 (2020)
4. Sulong, A., Gaaz, T., Sahari, J.: Mechanical and physical properties of injection molded halloysite nanotubes-thermoplastic polyurethane nanocomposites. Procedia Soc. Behav. Sci. **195**, 2748–2752 (2015)
5. Park, S.: Robust Design and Analysis for Quality Engineering. Chapman & Hall, London (1996)
6. Chie, S., Mansor, N., Keat, Y., Ai, T., Leng, T., Wei, K.: Effect of printing parameters on characteristic and properties PLA/Fe_2O_3 and CPLA/Fe_2O_3 composites using FDM techniques. Paper presented at AIP Conference Proceedings, vol. 2541, p. 050005 (2022)
7. Vidakis, N., Petousis, M., Korlos, A., Velidakis, E., Mountakis, N., Charou, C., Myftari, A.: Strain rate sensitivity of polycarbonate and thermoplastic polyurethane for various 3D printing temperatures and layer heights. Polymers **13**, 2752 (2021)
8. Kandi, R., Pandey, P.: Statistical modelling and optimization of print quality and mechanical properties of customized tubular scaffolds fabricated using solvent-based extrusion 3D printing process. Paper presented at proceedings of the institution mechanical engineers, vol. 235, pp. 1421–1438 (2021) (Part H: Journal of Engineering in Medicine)
9. Wang, J., Yang, B., Lin, X., Gao, L., Liu, T., Lu, Y., Wang, R.: Research of TPU materials for 3D printing aiming at non-pneumatic tires by FDM method. Polymers **12**, 2492 (2020)
10. Le, D., Nguyen, C., Pham, T., Nguyen, V., Pham, S., Le, M., Nguyen, T.: Optimizing 3D printing process parameters for the tensile strength of thermoplastic polyurethane plastic. J. Mater. Eng. Perform. **32**, 10805–10816 (2023)
11. Mohamed, S., Tirkes, S., Akar, A., Tayfun, U.: Hybrid nanocomposites of elastomeric polyurethane containing halloysite nanotubes and POSS nanoparticles: tensile, hardness, damping and abrasion performance. Clay Miner. **55**, 281–292 (2020)
12. Kwiatkowski, K., Nachman, M.: The abrasive wear resistance of the segmented linear polyurethane elastomers based on a variety of polyols as soft segments. Polymers **9**, 705 (2017)

13. Byrill, K., Piesowicz, E., Szymanski, P., Slaczka, W., Pijanowski, M.: Polymer composite manufacturing by FDM 3D printing technology. Paper presented at MATEC web of conferences, vol. 237, p. 02006 (2018)
14. Ozdil, N., Kayseri, G., Menguc, G.: Chapter 7: Analysis of abrasion characteristics in textiles. In: Adamiak M (ed.) Abrasion Resistance of Materials. IntechOpen Limited, London (2012)
15. Beck, R., Truss, R.: Effect of chemical structure on the wear behaviour of polyurethane-urea elastomers. Wear **218**, 145–152 (1998)
16. Hanief, M., Charoo, M.: Archard's wear law revisited to measure accurate wear coefficient considering actual sliding velocity. Mater. Today Proc. **47**, 5598–5600 (2021)
17. Anisimov, V., Semenets, A., Letunovskii, M., Strakhov, V.: Effect of rigid blocks on the mechanical characteristics and abrasive resistance of polyurethanes. Mater. Sci. **38**, 95–98 (2002)
18. Ryszkowska, J.: Supermolecular structure, morphology and physical properties of urea-urethane elastomers. Polimery **57**, 775–902 (2012)
19. Nachman, M., Kwiatkowski, K.: The effect of thermal annealing on the abrasion resistance of a segmented block copolymer urethane elastomers. Wear **306**, 113–118 (2013)
20. Michalski, P., Nachman, M., Kwiatkowski, K.: Abrasive wear of urethane elastomers based on monomeric or polymeric diisocyanate MDI. Polimery **57**, 839–845 (2012)
21. Namathoti, S., Vakkalagadda, M.: Mechanical and shape recovery characterization of MWCNTs/HNTs-reinforced thermal-responsive shape-memory polymer nanocomposites. Polymers **15**, 710 (2023)
22. Mohammadzadeh, F., Asl, V., Balzade, Z., Jouibari, I.: Switch segment and halloysite nanotube role in the phase separation behavior of shape-memory thermoplastic polyurethane. Polym. Compos. **41**, 1–9 (2020)
23. Li, X., Nikiforow, I., Pohl, K., Adams, J., Johannsmann, D.: Polyurethane coatings reinforced by halloysite nanotubes. Coatings **3**, 16–25 (2013)
24. Luo, P., Zhao, Y., Zhang, B., Liu, J., Yang, Y., Liu, J.: Study on the adsorption of Neutral Red from aqueous solution onto halloysite nanotubes. Water Res. **44**, 1489–1497 (2010)
25. Maamoun, A., Elkhateeb, A., Zulfiqar, S.: Halloysite-decorated mechanically robust polyurethane nanocomposite foams for acoustic relevance. Ind. Eng. Chem. Res. **61**, 17937–17949 (2022)
26. Carriço, C., Fraga, T., Carvalho, V., Pasa, V.: Polyurethane foams for thermal insulation uses produced from castor oil and crude glycerol biopolyols. Molecules **22**, 1091 (2017)
27. Chen, J., Tsai, M., Tseng, I., Hsu, A., Liu, T., Huang, S.: Composition, thermal and tensile properties of polyurethane-urea-silica hybrids. RSC Adv. **3**, 9729–9738 (2013)
28. Petrovic, Z., Zavargo, Z., Flyn, J., Macknight, W.: Thermal degradation of segmental polyurethanes. J. Appl. Polym. Sci. **51**, 1087–1095 (1994)
29. Wu, W., Zhao, W., Gong, X., Sun, Q., Cao, X., Su, Y., Yu, B., Li, R., Vellaisamy, R.: Surface decoration of halloysite nanotubes with POSS for fire-safe thermoplastic polyurethane nanocomposites. J. Mater. Sci. Technol. **101**, 107–117 (2022)
30. Du, M., Guo, B., Jia, D.: Thermal stability and flame retardant effects of halloysite nanotubes on poly (propylene). Eur. Polym. J. **42**, 1362–1369 (2006)
31. Jouibari, I., Asl, V., Ahmadi, H., Mirhosseini, M.: Micro-phase separation kinetics of polyurethane nanocomposites with neural network. Polym. Compos. **40**, 3904–3913 (2019)
32. Razzaq, M., Anhalt, M., Frormann, L., Weidenfeller, B.: Mechanical spectroscopy of magnetite fillet polyurethane shape memory polymers. Mater. Sci. Eng. A **471**, 57–62 (2007)
33. Mamaqani, H., Asl, V., Kalajahi, M.: In situ controlled radical polymerization: a review on synthesis of well-defied nanocomposites. Polym. Rev. **52**, 142–188 (2012)

Chapter 10
Influence of HNT Content and 3D Printing Parameters on Shape Memory Behaviour and Compressive Properties of PU/HNT Nanocomposites

Abstract This chapter evaluated the effects of different halloysite nanotube (HNT) content on shape memory behaviour and compressive properties of polyurethane (PU)/HNT nanocomposites manufactured via fused deposition modelling (FDM). Strip-like and re-entrant (RE) structures were manufactured in this study based on PU (MM4520 grade) and HNTs with the combination of solution casting, melt extrusion and 3D printing. Such samples were additively manufactured on the basis of optimum factor-level combination for L_{18} orthogonal array (OA) in response to maximum tensile strength at yield.

Keyword PU/HNT nanocomposites · Re-entrant structures · Shape memory behaviour · Compressive properties

10.1 Compressive Tests

Compressive tests were initially carried out at room temperature for 3D printed PU/HNT nanocomposites at different HNT contents of 0, 2, 4, 6, 8 and 10 wt% in terms of compressive strength, compressive modulus, specific energy absorption (SEA) and Poisson's ratio (PR). Corresponding structural models and dimensions, as well as experimental setup in compressive tests were aforementioned in Fig. 6.11a and b respectively. Nominal stress, nominal strain, energy absorption efficiency, SEA, PR of RE samples were determined according to corresponding Eqs. (6.3)–(6.5), (6.7) and (6.11). Relevant analytical methods were detailed in Sect. 6.3.3 to calculate compressive properties of resulting PU/HNT nanocomposites.

The original version of the chapter has been revised. A correction to this chapter can be found at https://doi.org/10.1007/978-981-95-0541-8_12

W. T. Nugroho et al., *Thermoplastic Polyurethane (TPU)/Halloysite Nanotube (HNT) Nanocomposites*, Engineering Materials, https://doi.org/10.1007/978-981-95-0541-8_10

10.2 Shape Memory Tests

As mentioned earlier, shape memory tests consisted of preliminary and main tests. In the preliminary tests, a three-point bending method was employed, which was followed by a material recovery process in order to determine specific temperature to heat up PU/HNT nanocomposite samples in compression and recovery processes. Furthermore, resulting temperature data was used in subsequent main tests. Strip-like nanocomposite samples according to ASTM D790-17 standard were chosen in preliminary tests under compression on a universal testing machine UTM-25 (IPC Global, Milan, Italy) with the aid of a mounted environmental chamber. Compressive tests were conducted at the crosshead speed of 1 mm/min with the load cell capacity of 25 kN. Whereas three-point bending tests were also carried out before and after loading on UTM-25 testing machine depicted in Fig. 6.13.

In shape memory tests (Fig. 10.1), nanocomposite samples were initially heated up to 50 °C and further underwent compression, which meant to be in the working condition above glass transition temperature (T_g) of PU. Afterwards, compressive load was maintained, while nanocomposite samples were cooled to room temperature when the load was removed. After dimensional measurements of nanocomposite samples, they were then heated up to specific temperatures of 60, 70 and 80 °C in this study until original shapes were regained. Effective heating temperatures in heating and recovery processes were obtained, and then further used in main tests. RE nanocomposite samples were tested on an asphalt mixture performance tester (AMPT) at the crosshead speed of 1 mm/min with the load capacity of 15 kN. Further, shape recovery ratio (R_{rs}) of strip-like nanocomposite samples, shape fixity ratio (R_f) of RE nanocomposite samples, as well as shape recovery ratio (R_r) of RE samples were then determined according to Eqs. (6.13)–(6.15) with detailed analytical methods aforementioned in Sect. 6.3.4.

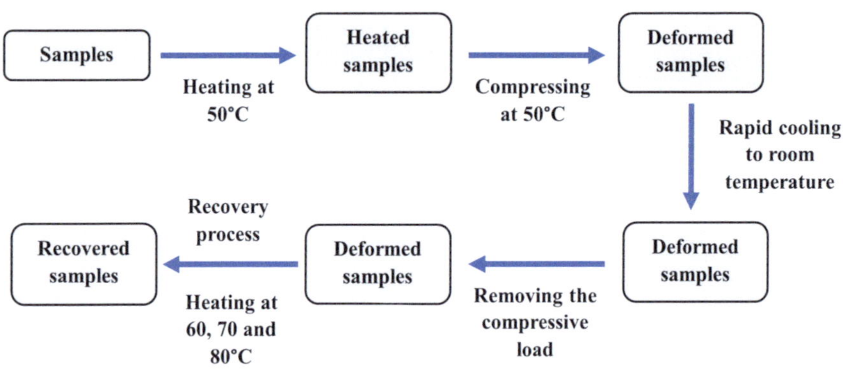

Fig. 10.1 Flowchart of shape memory tests on PU/HNT nanocomposites

10.3 Shape Memory Behaviour of PU/HNT Nanocomposites

Microphase separation involving both hard and soft segments in shape memory polymers generally influences their shape memory behaviour [1]. In the programming stage, soft segments fix deformed hard segments after bending nanocomposite samples at 50 °C (above their T_g) before cooling to room temperature (below their T_g). Resulting stress in this process enables to be stored between cross-linked hard segments and released after being heated in the recovery process.

In general, HNTs tend to absorb hard segments located in soft segment region in order to induce the purification of soft segments where soft segments are key to shape recovery and original shape configuration [2]. HNT aggregation at higher contents inevitably restricts the mobility of hard segments leading to the reduction of material flexibility [3]. Hence, deformed nanocomposite samples were prevented from retrieving their original shapes in the recovery process.

Figure 10.2 illustrates the mechanism of using specific load and heat stimuli to achieve shape recovery of strip-like nanocomposite samples. Their chemical structures consist of both hard and soft segments shown in Fig. 10.2a. Further, they were deformed in three-point bending tests with applied load placed in the vertical direction of strip-like samples. Then such samples were heated in an oven after bending deformation. Under this condition, heat transfer took place to mobilise hard segments in order to resume their original shapes. As seen in Fig. 10.2b, the presence of HNTs within PU matrices in nanocomposite samples hindered the mobility of hard segments to regain their initial shapes.

Shape memory behaviour of nanocomposites at different HNT contents was depicted in Fig. 10.3. Tr-60C, Tr-70C and Tr-80C were referred to as three different temperatures (T_r) at 60, 70 and 80 °C. Apparently, the inclusion of HNTs prevented deformed nanocomposite samples from shape recovery. Increasing HNT content

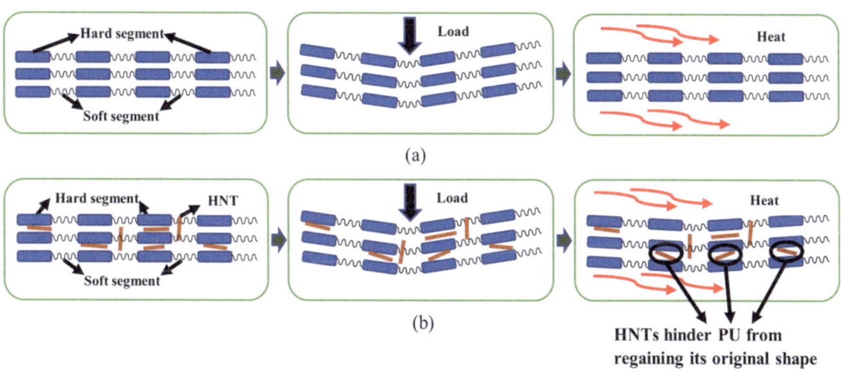

Fig. 10.2 Schematic diagram of deformation process for **a** neat PU and **b** PU/HNT nanocomposites

decreased recovery ratio according to Fig. 10.3a in good accordance with Mohammadzadeh et al. [2]. On the contrary, Namathoti et al. [4] reported that increasing HNT content enhanced shape recovery ratio, but decreased recovery time. This finding may be attributed to different fabrication processes (i.e. 3D printing versus twin screw extrusion and injection moulding) applied along with different processing conditions and HNT contents.

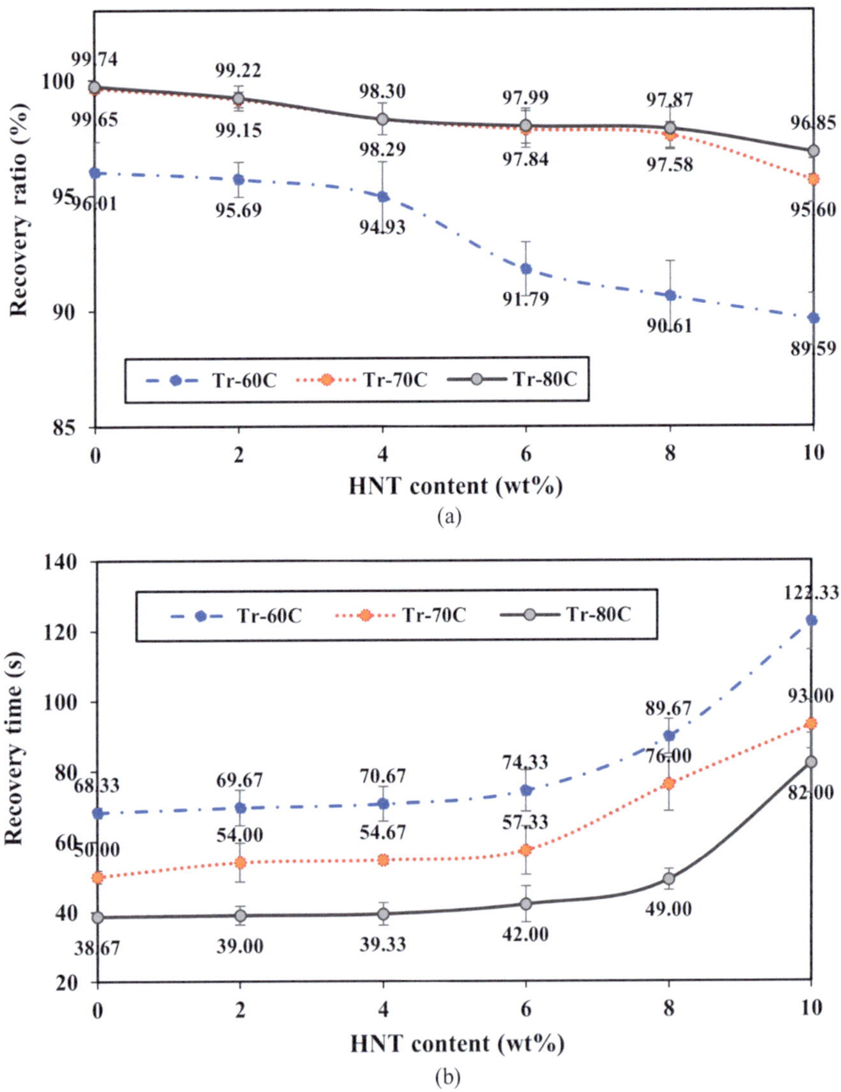

Fig. 10.3 **a** Recovery ratio and **b** recovery time of nanocomposite samples based on the optimum factor-level combination in response to maximum tensile strength at yield

It is expected that recovery temperature enables to significantly affect recovery ratio. Higher T_r generally yields a greater recovery ratio in relation to heat transfer from a heating source like an oven to PU/HNT nanocomposites. The heat allows for the mobility of hard segments for shape recovery after deformation. However, heat energy was inadequate for complete recovery of nanocomposite samples at 100% when heated to 60 °C (i.e. Tr-60C). At this temperature level, the greatest recovery ratio of neat PU was only achieved at 96.01% with a further gradually declining trend at higher HNT contents in nanocomposite samples. The sharp drop in recovery ratio took place at the HNT content of 6 wt%, which can be ascribed to HNT aggregation to obstruct hard segments of PU matrices for shape recovery. At the highest HNT content of 10 wt%, the minimum recovery ratio of 89.59% was achieved in nanocomposite samples.

It was evidently shown in Fig. 10.3a that increasing recovery temperature up to 70 °C (i.e. Tr-70C) yielded recovery ratios of 99.65% and 95.60% for neat PU and nanocomposite samples at the HNT content of 10 wt%, respectively. More evidently, heat energy supplied from the oven facilitated the mobility of PU molecular chains in nanocomposite samples leading to better shape recovery than that for Tr-60C. A further increase of temperature to 80 °C (i.e. Tr-80C) slightly improved recovery ratio, especially for nanocomposite samples reinforced with 10 wt% HNTs. Nonetheless, as for Tr-70C, insufficient thermal energy and HNT aggregation resulted in the hindrance of PU shape recovery.

Figure 10.3b presented the effect of HNT content and recovery temperature on recovery time of nanocomposite samples at different HNT contents. Increasing HNT content tended to prolong recovery time due to mobility restriction of hard segments within PU matrices in nanocomposite samples. As such, more recovery time could be required for corresponding nanocomposite samples at higher HNT content levels. The longest recovery time of 122 s was identified for nanocomposite samples reinforced with 10 wt% HNTs, as opposed to the shortest recovery time of 68 s for neat PU. When increasing T_r up to 70 °C, recovery time of nanocomposite samples reinforced with 10 wt% HNTs and neat PU became 93 and 50 s. A further temperature increase up to 80 °C reduced recovery time of corresponding nanocomposites reinforced with 10 wt% and neat PU to 82 and 37 s respectively, which could be associated with the amount of heat transfer required by PU for shape recovery. The higher recovery temperature induced greater energy to accelerate shape recovery process, thereby resulting in shorter recovery time. It could be concluded that effective heating temperatures in three-point bending tests and recovery process were 50 and 80 °C accordingly, which were then employed for main shape memory tests of RE structures and 3D printed samples.

RE nanocomposite samples had different HNT contents in range of 0–10 wt% similar to strip-like samples. However, RE samples were deformed under three-point bending rather than compression. Moreover, heating temperatures of 50 and 80 °C were applied for both compression and recovery processes, thus resulting in shape fixity ratio (R_f), shape recovery ratio (R_r) and recovery time. As mentioned earlier, the incorporation of HNTs decreased the mobility of hard segments and prevented deformed nanocomposite samples from regaining their original shape. On the other hand, R_f reveals the ability to switch different segments to lock in transient compressed height h_h upon the load release. Under this circumstance, most shape memory materials can be instantaneously stretched from h_h to h_i. In particular, R_f becomes 100% when $h_i = h_h$ according to Eq. (6.14), as illustrated in Fig. 6.15.

Temporary shape of RE nanocomposite samples can be locked in through strain-induced crystallisation of soft segments to reduce switching transition temperature [1]. After heating nanocomposite samples above T_g, as well as further compressed and cooled below T_g, polymeric chains of nanocomposite samples appeared to maintain their deformed condition by soft segments. In addition to crystallites derived from hard segments, the crystallisation with respect to soft segments is induced by the deformation in the rubbery state at the cooling step, which further restricts the mobility of polymeric chains, as well as strain recovery with the absence of applied stress. As such, when the load is not considered, the retention of temporary shape should be achieved. Nonetheless, R_f of PU RE samples reached approximately 80.21% illustrated in Fig. 10.4a, which implied that deformed height was unable to be retained upon the load release. When the load was removed, RE nanocomposite samples were instantaneously stretched from h_h to h_i. This indicated that soft segments provided enough impediment to the relaxation of deformed segments due to their partial crystallisation.

The incorporation of HNTs increased R_f of RE nanocomposite samples owing to the effect of strain-induced crystallisation [5], which was the case when considering the role of HNTs as rigid nanofillers to promote the crystallinity of soft segments. On the other hand, strain distribution within PU matrices adjacent to HNTs further expanded in the deformation process, which could be induced by significant modulus difference between rubber-like matrices and HNTs [5]. Strain localisation yielded the increase in strain-induced crystallisation arising from a greater degree of crystallinity. In this way, the relaxation of deformed polymeric chains was precluded accordingly. Increasing HNT content improved strain-induced crystallisation, as evidenced by lower h_i leading to higher R_f.

The capability of shape memory materials to recover their initial shape is generally represented by R_r. RE nanocomposite samples were heated to recover their original height h_0 from temporary height h_i. In particular, certain shape memory materials enable to only regain their length up to h_e. At the recovery step, internal stresses between the crystallites of physically cross-linked hard and soft segments stored in the deformation process could be released when RE nanocomposite samples were heated above their T_g. As such, this reduced the rigidity of soft segments, and further improved micro-Brownian mobility [5]. The frozen stress was active at the shape recovery step. It is well understood that shape recovery depends primarily on the capability of hard segments in order to maintain initial shape via inter- or intra-polymeric chain attraction, also called hydrogen bonding or dipole–dipole interaction. The inclusion of HNTs reduced the mobility of polymeric chains, and increasing HNT content inevitably decreased R_r.

The inclusion of HNTs as rigid nanofillers tended to restrict the mobility of polymeric chains in hard segments of PU molecules. Therefore, longer recovery time was often required for RE nanocomposites samples with higher HNT contents in order

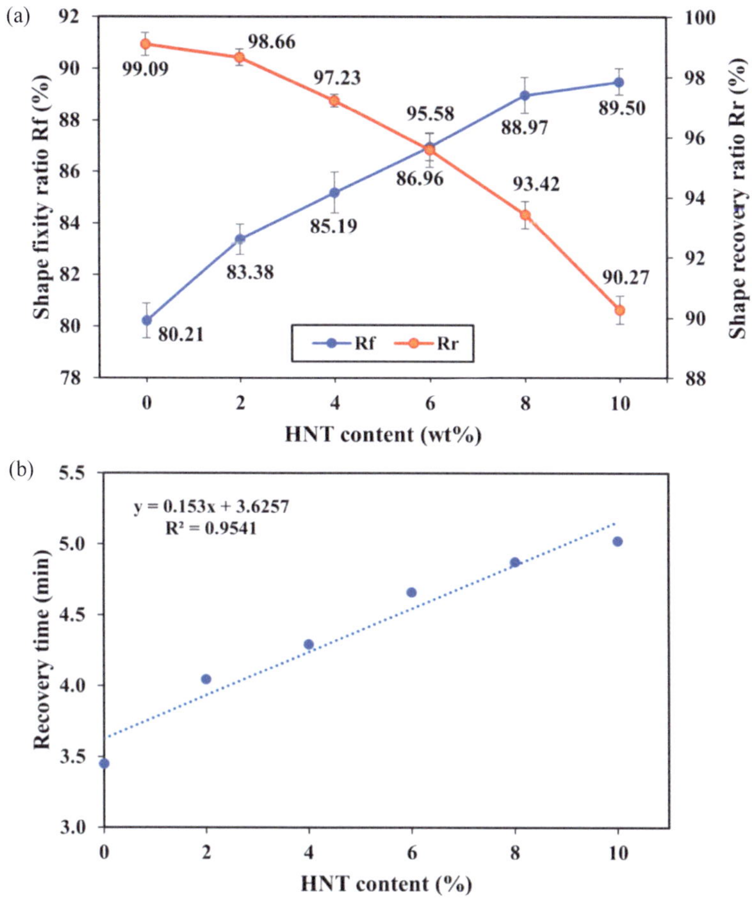

Fig. 10.4 **a** Shape fixity ratio and shape recovery ratio and **b** recovery time for RE nanocomposite samples

to recover their initial shape according to Fig. 10.4b. In contrast, RE nanocomposite samples had more complicated geometries when compared with strip-like samples, particularly in the curved regions. This led to longer recovery time and lower shape recovery ratio. Recovery time increased from 3.45 to 5.02 min in a monotonic manner when increasing the HNT content from 2 to 10 wt%. Typical shape memory behaviour of RE nanocomposite samples were presented in Table 10.1.

Table 10.1 Shape memory behaviour of RE nanocomposite samples

HNT content (wt%)	Shape fixity ratio R_f (%)	Shape recovery ratio R_r (%)	Recovery time (min)
0	80.21	99.09	3.45
2	83.38	98.66	4.04
4	85.19	97.23	4.29
6	86.96	95.58	4.66
8	88.97	93.42	4.87
10	89.50	90.27	5.02

10.4 Compressive Properties of PU/HNT Nanocomposites

Compressive properties are deemed a critical indicator to evaluate mechanical properties of RE nanocomposite samples. Figure 10.5 exhibited a typical stress–strain curve under compression, which consists of elastic, plastic and densification regions. In particular, elastic region at the curve state demonstrated a linearly elastic behaviour of materials where elastic bending of cell structs prevailed. Accordingly, compressive modulus was determined from elastic deformation regime. The second plateau region is known as plastic region with stress remaining unchanged with increasing the strain (or deformation) of cell struts. The third densification region referred to significantly increasing the stress with increasing the strain for RE nanocomposite samples. The skewed curve portion in the first elastic region was targeted to identify the modulus of materials, also known as their stiffness (i.e. resistance to deformation). SEA was also calculated by dividing the region under the curve portion from 0 to onset strain of densification ε_d with the involvement of the density of RE nanocomposite samples [6]. Before the determination of ε_d, energy absorption efficiency $\eta(\varepsilon)$ was determined initially according to Eq. (6.5). ε_d took place when $\eta(\varepsilon)$ reached the maximum level. Compressive strength of RE nanocomposite samples was indicated by the highest curve peak identified prior to the densification stage. It was aforementioned that PR of RE nanocomposite sampled could be defined by using Eq. (6.11), which was determined based on 20 data points with respect to compressive strain collected ranging from 0 to ε_d.

Figure 10.5a revealed typical stress–strain curves of RE nanocomposites under compression at the HNT contents ranging from 0 to 10 wt%, which were referred to as PU, NC-2wt%HNT, NC-4wt%HNT, NC-6wt%HNT, NC-8wt%HNT and NC-10wt%HNT respectively. Apparently increasing HNT content mostly enhanced compressive curves to higher stress levels, which implied positive impact of embedded nanofillers to improve compressive properties of RE samples. Overall, ε_d of RE nanocomposite samples appeared to be smaller when compared with that of neat PU. It was worth noting from Fig. 10.5b that RE nanocomposite samples reinforced with 10 wt% HNTs possessed ε_d up to 38.80% relative to 44.52% for neat PU. This finding might suggest that the incorporation of HNTs increased the stiffness of RE nanocomposite samples leading to accelerated material densification.

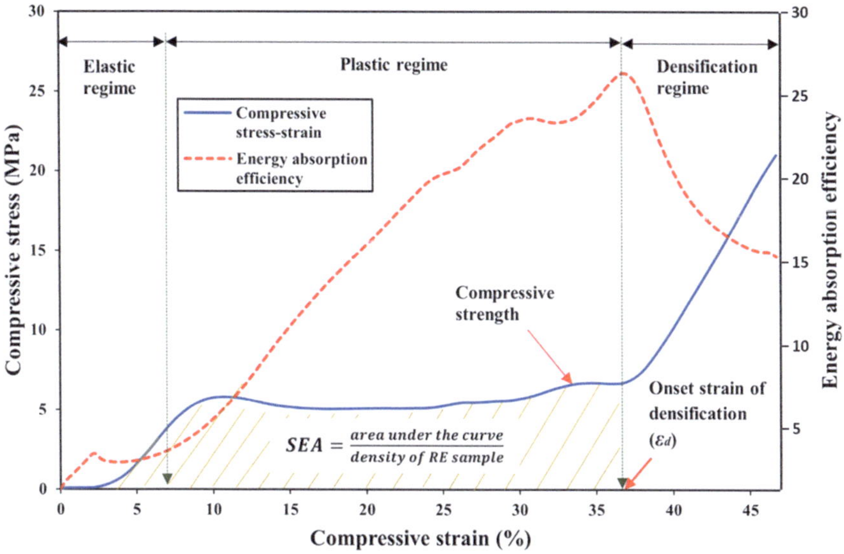

Fig. 10.5 A typical compressive stress-compressive strain curve with corresponding energy absorption efficiency

Typical trendline functions with respect to stress–strain curves under compression for RE nanocomposite samples at different HNT contents, as well as corresponding integration results were presented in Tables 10.2 and 10.3. All trendline functions had R^2 values over 0.95, indicating the variation of compressive stress depended on compressive strain as an independent variable with good curve fitting. Subsequently, these trendline functions were integrated with the aid of MATLAB R2023a software, as earlier mentioned in Sect. 6.3.3.

The integration results of these trendline functions were employed to determine energy absorption efficiency according to Eq. (6.5). Furthermore, such integration results were used to calculate SEA based on Eq. (6.7) with required density values of RE nanocomposite samples. The densities of RE samples were determined using density kits according to Eqs. (6.1) and (6.8), as listed in Table 10.4. Neat PU employed in this study has an average density of 1200 kg/m^3, which is slightly lighter than that mentioned by the supplier at 1250 kg/m^3 [7]. It was clearly observed that increasing HNT content up to 10 wt% induced trivial effect to improve the density of RE nanocomposite samples by only 1.83%. On the other hand, the impact of HNT inclusion on compressive strength, compressive modulus and SEA of RE nanocomposite samples were also presented in Fig. 10.6a and b.

The incorporation of HNTs within PU matrices in RE nanocomposite samples enhanced their compressive strength, particularly at the HNT content of 4 wt%, with a clear sign of good reinforcing mechanism of HNTs. A further increase in HNT content slightly promoted the compressive strength, and the highest compressive strength of RE nanocomposite samples was achieved at 5.20 MPa when reinforcing

Table 10.2 Best-fitting trendline functions of stress–strain curves under compression for RE nanocomposite samples

Sample	Trendline function	R^2
PU	$\sigma(\varepsilon) = 8 \times 10^{-7}\,\varepsilon^{5} - 910^{-5}\,\varepsilon^{4} + 3.9 \times 10^{-3}\,\varepsilon^{3} - 7.42 \times 10^{-2}\,\varepsilon^{2} + 0.6982\,\varepsilon - 0.7198$	0.99
NC-2wt%HNT	$\sigma(\varepsilon) = 3 \times 10^{-7}\,\varepsilon^{5} - 2 \times 10^{-5}\,\varepsilon^{4} + 6 \times 10^{-4}\,\varepsilon^{3} - 6.3 \times 10^{-3}\,\varepsilon^{2} + 0.202\,\varepsilon + 0.3514$	0.98
NC-4wt%HNT	$\sigma(\varepsilon) = 1 \times 10^{-6}\,\varepsilon^{5} - 1 \times 10^{-4}\,\varepsilon^{4} + 4.57 \times 10^{-3}\,\varepsilon^{3} - 9.2 \times 10^{-2}\,\varepsilon^{2} + 0.8067\,\varepsilon - 0.4499$	0.98
NC-6wt%HNT	$\sigma(\varepsilon) = 1 \times 10^{-6}\,\varepsilon^{5} - 1 \times 10^{-4}\,\varepsilon^{4} + 4.2 \times 10^{-3}\,\varepsilon^{3} - 7.25 \times 10^{-2}\,\varepsilon^{2} + 0.6331\,\varepsilon - 0.1783$	0.98
NC-8wt%HNT	$\sigma(\varepsilon) = 2 \times 10^{-6}\,\varepsilon^{5} - 1.735 \times 10^{-4}\,\varepsilon^{4} + 6.4 \times 10^{-3}\,\varepsilon^{3} - 0.1043\,\varepsilon^{2} + 0.8216\,\varepsilon - 0.2194$	0.98
NC-10wt%HNT	$\sigma(\varepsilon) = 2 \times 10^{-6}\,\varepsilon^{5} - 1.92 \times 10^{-4}\,\varepsilon^{4} + 7.5 \times 10^{-3}\,\varepsilon^{3} - 0.1213\,\varepsilon^{2} + 0.8986\,\varepsilon + 0.0102$	0.99

Note σ and ε denote compressive stress and compressive strain. σ (ε) is referred to as compressive stress function in terms of compressive strain

Table 10.3 Summary of integration results of trendline functions

Sample	Integration result of trendline function $F(x)$
PU	$\sigma'(\varepsilon) = 1.33 \times 10^{-7}\,\varepsilon^{6} - 1.810^{-5}\,\varepsilon^{5} + 0.975 \times 10^{-3}\,\varepsilon^{4} - 2.473 \times 10^{-2}\,\varepsilon^{3} + 0.3491\,\varepsilon^{2} - 0.7198\,\varepsilon + 0.2781$
NC-2wt%HNT	$\sigma'(\varepsilon) = 0.5 \times 10^{-7}\,\varepsilon^{6} - 0.4 \times 10^{-5}\,\varepsilon^{5} + 1.5 \times 10^{-4}\,\varepsilon^{4} - 2.1 \times 10^{-3}\,x^{3} + 0.101\,\varepsilon^{2} + 0.3514\,\varepsilon + 0.1057$
NC-4wt%HNT	$\sigma'(\varepsilon) = 0.167 \times 10^{-6}\,\varepsilon^{6} - 0.2 \times 10^{-4}\,\varepsilon^{5} + 1.1425 \times 10^{-3}\,\varepsilon^{4} - 3.067 \times 10^{-2}\,\varepsilon^{3} + 4.0335 \times 10^{-2}\,\varepsilon^{2} - 0.4499\,\varepsilon + 0.1253$
NC-6wt%HNT	$\sigma'(\varepsilon) = 0.167 \times 10^{-6}\,\varepsilon^{6} - 0.2 \times 10^{-4}\,\varepsilon^{5} + 1.04 \times 10^{-3}\,\varepsilon^{4} - 2.4167 \times 10^{-2}\,\varepsilon^{3} + 0.3165\,\varepsilon^{2} - 0.1783\,\varepsilon + 0.0192$
NC-8wt%HNT	$\sigma'(\varepsilon) = 0.33 \times 10^{-6}\,\varepsilon^{6} - 0.347 \times 10^{-4}\,\varepsilon^{5} + 1.6 \times 10^{-3}\,\varepsilon^{4} - 0.035x^{3} + 0.4108\,\varepsilon^{2} - 0.2194\,\varepsilon + 0.0221$
NC-10wt%HNT	$\sigma'(\varepsilon) = 0.33 \times 10^{-6}\,\varepsilon^{6} - 0.384 \times 10^{-4}\,\varepsilon^{5} + 1.875 \times 10^{-3}\,\varepsilon^{4} - 0.0404x^{3} + 0.4493\,\varepsilon^{2} + 0.0102\,\varepsilon + 0.1018$

Note σ and ε denote compressive stress and compressive strain. $\sigma'(\varepsilon)$ is referred to as their corresponding derivative function

Table 10.4 Density of RE nanocomposite samples

Sample	Density of PU ρ_s (kg/m^3)	Density of RE sample ρ (kg/m^3)
PU	1200	467.49
NC-2wt%HNT	1205	469.43
NC-4wt%HNT	1210	471.38
NC-6wt%HNT	1215	473.33
NC-8wt%HNT	1218	474.50
NC-10wt%HNT	1222	476.06

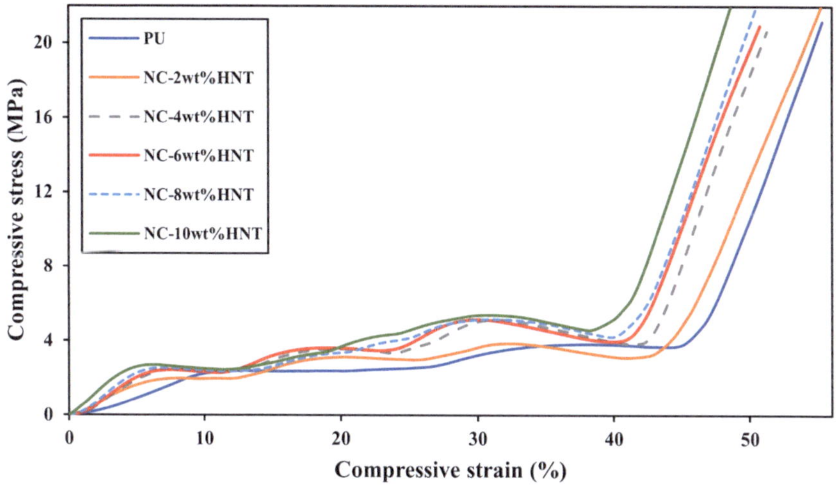

(a)

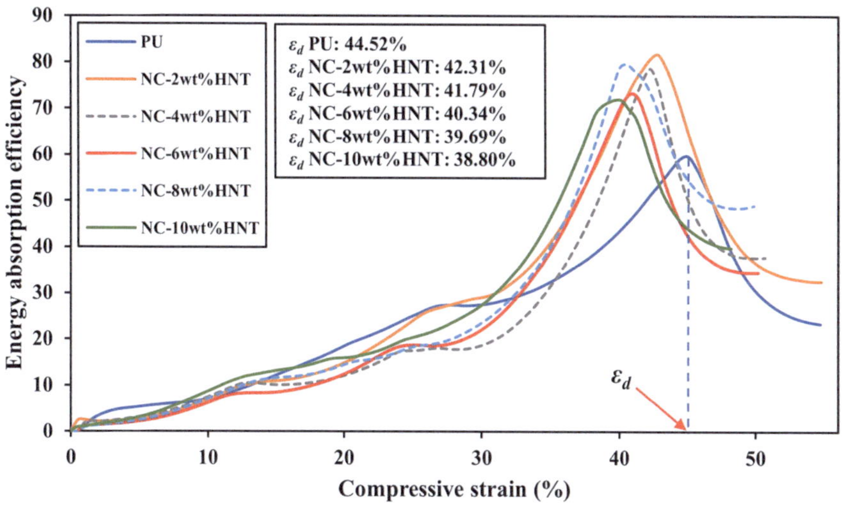

(b)

Fig. 10.6 **a** Stress–strain curves under compression and **b** energy absorption efficiency for RE nanocomposite samples

10 wt% HNTs, as opposed to that of neat PU at 3.88 MPa. Nie et al. [8] reported that compressive strength of polydopamine (PDA)-coated PU sponge called PDA-PUS at 50% strain was approximately 1.26 kPa, which was over four folds than that of PU sponge. Such a high increase was related to covalent bonding of PDA-PUS with γ-aminopropyltriethoxysilane (APTES)-grafted HNTS (A-HNTs), which led to the improvement in the rigidity of sponge skeleton. Lorusso et al. [9] reported that the inclusion of HNTs up to 10 wt% within PU matrices could enhance compressive

strength of foam samples from 96.84 MPa for neat PU to 106.56 MPa for corresponding nanocomposites. Increasing HNT content appeared to enhance compressive modulus of RE nanocomposite samples in a monotonic manner according to Fig. 10.7a. The presence of HNTs as rigid nanofillers within PU matrices gave rise to a higher degree of phase separation due to hydrogen bonding of urethane groups in PU hard segments with hydroxyl groups exposed to HNT surfaces [10]. This phenomenon resulted in a higher density of hydrogen bonds in hard segments to hinder the mobility of polymeric chains, thus resulting in higher stiffness [11]. In particular, the compressive modulus of RE nanocomposite samples with the addition of 10 wt% HNTs was significantly increased by a factor of 1.26 relative to that of neat PU according to Fig. 10.7a. Salaman et al. [12] revealed that the inclusion of 3 wt% HNTs within PU matrices in RE nanocomposite samples enhanced their compressive modulus to 1.67 GPa when compared with that of neat PU at 1.29 GPa.

Similar to compressive properties of RE nanocomposite samples, the inclusion of HNTs within PU matrices increased their density, which allowed these nanocomposite samples to sustain a greater load than neat PU. As such, increasing HNT content also enhanced SEA of RE nanocomposite samples accordingly, as illustrated in Fig. 10.7b. Maximum SEA of 0.76 kJ/kg was obtained at the HNT content of 10 wt% when compared with 0.44 kJ/kg for neat PU in good accordance with El-baky et al. [13] where HNTs were used as nanofillers to improve crashworthiness ability of tubular samples based on epoxy/glass composites. The incorporation of 1, 2, 3 and 4 wt% HNTs yielded significant increases of 160.61%, 215.30%, 190.96% and 233.91% for SEA of nanocomposite tubes, as opposed to unfilled tubes at 11.50 J/g. It might be induced by a toughening mechanism of HNTs with respect to front pinning, deflection and bridging of resulting cracks [13]. Once microcracks under a quasi-static axial load approached HNTs, such cracks became stabilised by a crack bridging mechanism of HNTs, which was in good agreement with Ye et al. [14] using HNTs to reinforce epoxy/carbon composites. Silva et al. [15] obtained similar results when embedding other nanofillers such as montmorillonite (MMT), nanosilica (SiO_2) and glass-sphere (GS) particles as the secondary reinforcements for polyamide 6 (PA6)/glass fibre (GF) composites. Overall, compressive properties of RE nanocomposite samples were listed in Table 10.5 in this study.

Figure 10.8 illustrated PR vs. compressive strain curves for RE nanocomposite samples in good accordance with previous study [6]. Here PR was determined according to lateral engineering strain and axial engineering strain when RE nanocomposite samples were under compression in the x-direction. The addition of HNTs within PU matrices hindered the mobility of polymeric chains leading to much higher stiffness. Finally, increasing HNT content was found to decrease engineering strain particularly in the lateral direction along with the reduction in negative Poisson's ratio (NPR).

RE nanocomposite samples had an auxetic characteristics due to their specific construction to allow them to undergo transverse contraction when subjected to uniaxial compression [16]. Once compression strain reached 10%, the curve regions of the structure in the middle area of RE nanocomposite samples touched each other along with a lower lateral strain, as evidenced by the curve change from a sharp

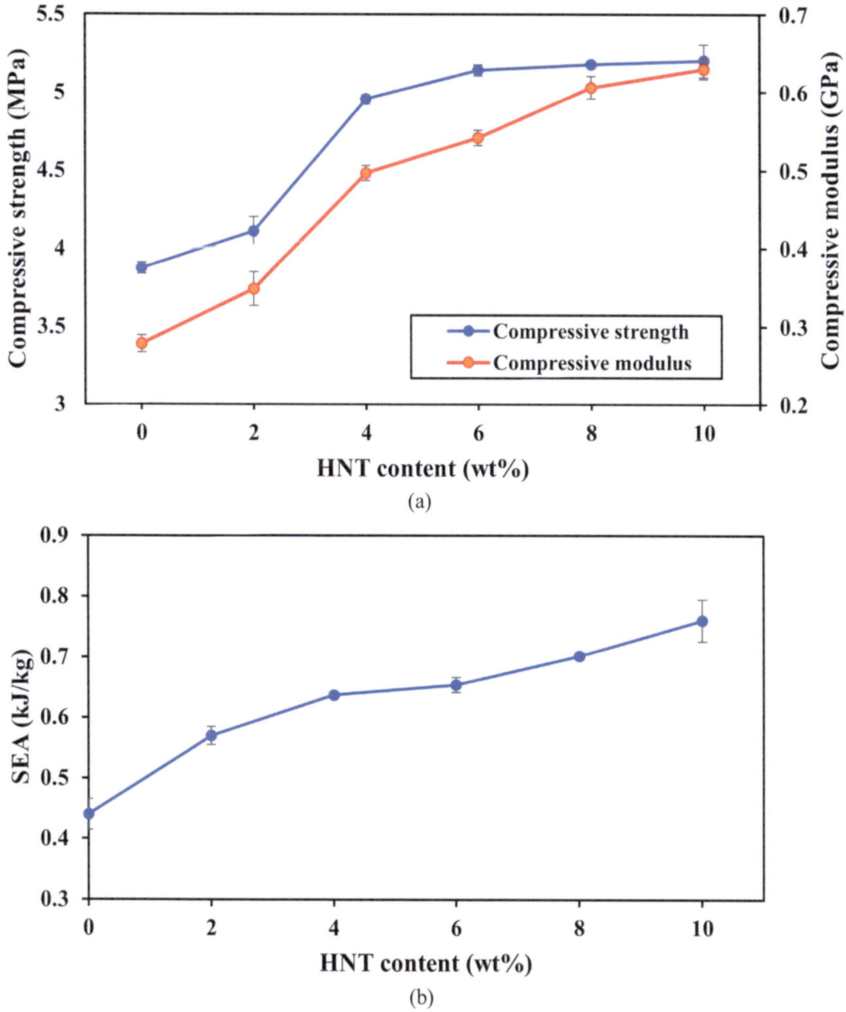

Fig. 10.7 Effect of HNT content on **a** compressive strength and compressive modulus and **b** SEA of RE nanocomposite samples

Table 10.5 Compressive properties of RE nanocomposite samples

HNT content (wt%)	Compressive strength (MPa)	Compressive modulus (GPa)	SEA (kJ/kg)
0	3.88	0.28	0.44
2	4.11	0.35	0.57
4	4.96	0.50	0.64
6	5.14	0.54	0.65
8	5.18	0.61	0.70
10	5.20	0.63	0.76

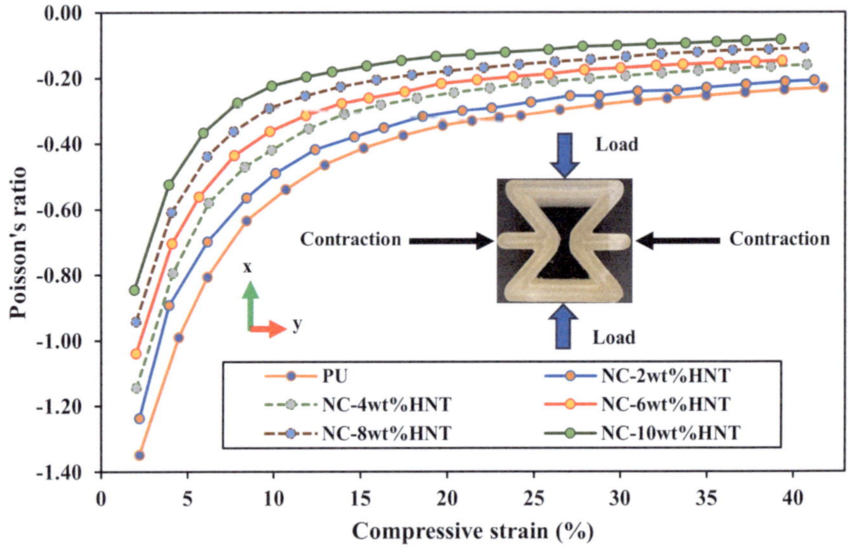

Fig. 10.8 Effect of HNT content on NPR of RE nanocomposite samples

increase to a gradual increase. This result coincided with Park et al. [17], which analysed the performance of chondrocyte proliferation within PU auxetic scaffolds under compression for the applications of articular cartilage in tissue engineering.

10.5 Summary

The effect of HNT content on compressive properties and shape memory behaviour of RE nanocomposite samples was studied in this chapter. The addition of 10 wt% HNTs within PU matrices in RE nanocomposite samples significantly improved their compressive strength, compressive modulus, as well as SEA by 34%, 125% and 73% respectively when compared with those of neat PU. Such results might be correlated with better interfacial bonding between surface functional groups of HNTs and hard segments of PU. In contrast, increasing HNT content reduced engineering strain, particularly lateral engineering strain, owing to the restriction to the mobility of polymeric chains to achieve a decrease in NPR.

As for shape memory behaviour, the presence of HNTs restricted the mobility of polymeric chains with three major existing effects. The first effect was related to the improved capability of PU/HNT nanocomposites at the programming step to enhance their shape fixity ratio. The inclusion of 10 wt% HNTs increased R_f by 12% for RE nanocomposite samples relative to that of neat PU. The second effect was associated with the restriction of HNTs to prevent PU matrices from recovering their original shape, thereby resulting in decreasing R_r. Moreover, R_r was slightly dropped by 9%. The last effect might be due to the contribution of existing HNTs to hinder PU from regaining its original shape, and further delay recovery time by 45% with the addition of 10 wt% HNTs.

References

1. Abdullah, S., Jumahat, A., Abdullah, N., Frormann, L.: Determination of shape fixity and shape recovery rate of carbon nanotube-filled shape memory polymer nanocomposites. Proc. Eng. **41**, 1641–1646 (2012)
2. Mohammadzadeh, F., Asl, V., Balzade, Z., Jouibari, I.: Switch segment and halloysite nanotube role in the phase separation behavior of shape-memory thermoplastic polyurethane. Polym. Compos. **41**, 1–9 (2020)
3. Sulong, A., Gaaz, T., Sahari, J.: Mechanical and physical properties of injection molded halloysite nanotubes-thermoplastic polyurethane nanocomposites. Proc. Soc. Behav. Sci. **195**, 2748–2752 (2015)
4. Namathoti, S., Vakkalagadda, M.: Mechanical and shape recovery characterization of MWCNTs/HNTs-reinforced thermal-responsive shape-memory polymer nanocomposites. Polymers **15**, 710 (2023)
5. Koerner, H., Price, G., Pearce, N., Alexander, M., Vaia, R.: Remotely actuated polymer nanocomposites–stress-recovery of carbon-nanotube-filled thermoplastic elastomers. Nat. Mater. **3**, 115–120 (2004)
6. Alomarah, A., Masood, S., Sbarski, I., Faisal, B., Gao, Z., Ruan, D.: Compressive properties of 3D printed auxetic structures: experimental and numerical studies. Virtual Phys. Prototyp. **15**, 1–21 (2019)
7. SMP Technologies: Shape Memory Polymer. SMP Technologies (2021). http://www2.smptechno.com/en/smp/. Accessed 5 Mar 2022
8. Nie, K., Ji, N., Xu, J., Guo, M., Wang, Y., Xu, X.: Fast and efficient removal of dyes and Cr(VI) ion from wastewater using halloysite nanotubes-covalently coated polyurethane sponges. J. Environ. Chem. Eng. **10**, 108308 (2022)
9. Lorusso, C., Vergaro, V., Conciauro, F., Ciccarella, G., Congedo, P.: Thermal and mechanical performance of rigid polyurethane foam added with commercial nanoparticles. Engineering **7** (2017)
10. Mahunaki, M., Asl, V., Mamaqani, H., Koosha, M., Yazdi, M.: Halloysite-reinforced thermoplastic polyurethane nanocomposites: physico-mechanical, rheological, and thermal investigations. Polym. Compos. **41**, 3260–3270 (2020)
11. Aguiar, R., Petel, O., Miller, R.: Effect of a halloysite-polyurethane nanocomposite interlayer on the ballistic performance of laminate transparent armour. Compos. Part C Open Access **7**, 100231 (2022)
12. Salaman, A., Obaidi, A., Takriff, M.: Enhancing morphology and compression properties of halloysite reinforced polyurethane nanocomposites using injection-moulding technique. Results Phys. **14**, 102507 (2019)
13. El-baky, M., Hegazy, D., Hassan, M., Kamel, M.: Potentiality of halloysite nanoclay on crashworthiness performance of polymer composite tubular elements. J. Compos. Mater. **56** (2022)
14. Ye, Y., Chen, H., Wu, J., Chan, C.H.: Evaluation on the thermal and mechanical properties of HNT-toughened epoxy/carbon fibre composites. Compos. B Eng. **42**, 2145–2150 (2011)
15. Silva, F., Sachse, S., Njuguna, J.: Energy absorption characteristics of nano-composite conical structures. Paper presented at IOP conference series: materials science and engineering, vol. 40, p. 012020 (2012)
16. Ozturk, M., Baran, T., Tatlier, M.: Experimental and numerical investigation of conventional and stiffened re-entrant cell structures under compression. J. Br. Mech. Eng. **44**, 593 (2022)
17. Park, Y., Kim, J.: The effect of negative Poisson's ratio polyurethane scaffolds for articular cartilage tissue engineering applications. Adv. Mater. Sci. Eng. **2013**, 853289 (2013)

Chapter 11
Applications

Abstract Chapter 11 focuses on potential applications of polyurethane (PU)/halloysite nanotube (HNT) nanocomposites via 3D printing in range of six different areas including biomedical engineering, electronic and antistatic devices, thermal insulation, sports equipment and garments for future product development.

Keyword PU/HNT nanocomposites · 3D printing · Applications

Special objects with complex geometry and voids or constructed from struts such as lattice structures, are very challenging to be manufactured by using conventional methods such as extrusion and injection moulding. Such an issue has been overcome due to the use of 3D printing technologies, enabling to build an object by depositing materials in a layer-by-layer manner. This unique characteristic attracts scientists, engineers and industrial manufacturers to utilise 3D printing in order to manufacture different types of prototypes and even end products. Meanwhile, PU with a shape memory behaviour is a good material candidate to be processed by 3D printing and create functional objects used for widespread applications ranging from biomedical engineering, electronics, sports equipment, etc.

11.1 Biomedical Applications

Reconstructive surgical operations have been carried out millions of times to tackle bone injury due to the destruction of bone tissues every year [1]. Bone tissue annihilation is induced by traumatic injury or after traumatic injury known as ineffective healing. Human bodies can heal bone breakage by themselves using distinct bone cells commencement with bleeding. When the bleeding occurs, the blood flows to the location of bone breakage, establishes a clot and generates swelling in fracture areas. The necessary time for healing the injuries may be dissimilar for everyone. It depends on health conditions, age, breakage type and blood supply condition

W. T. Nugroho et al., *Thermoplastic Polyurethane (TPU)/Halloysite Nanotube (HNT) Nanocomposites*, Engineering Materials, https://doi.org/10.1007/978-981-95-0541-8_11

including the presence of infection [2]. Small bone defects are typically healed naturally by means of renewal, growth and repair mechanisms. However, when large bone defects exceed a critical size, the healing process is of particular importance. Under this circumstance, bone regeneration based on tissue engineering methods is in high demand. Accordingly, scaffolds should be contained as an essential part of tissue engineering to target tissue regeneration like bone defects.

A scaffold model requires a high accuracy level, and its application in human bodies needs to match bone defect contour. PU composites with a shape memory behaviour could meet these stringent requirements. Such a unique property allows PU composites to maintain a temporary programmed stage and regain their initial form after combining a bone, as per requirement using minimal invasive bone repair [4]. PU, polylactic acid (PLA) and graphene oxide (GO) were selected to prepare nanocomposite scaffolds via fused deposition modelling (FDM) [3] according to Fig. 11.1.Those merged materials could increase thermal stability and mechanical strength of 3D printed scaffolds. Furthermore, these scaffolds possessed remarkable biocompatibility with NIH3T3 cells, which considered their critical role as potential scaffolds used for tissue engineering. Additionally, printing direction also influenced mechanical response of scaffolds due to poor adhesion between layers in 3D printing process as a typical disadvantage of FDM.

PU composites enable to be used to produce flexible sensors for prosthetics with the inclusion of carbon nanotubes (CNTs) as the nanofillers. Kim et al. [5] reported the manufacture of flexible pressure PU/multi-walled CNT (MWCNT) composite sensors via FDM. These sensors with the inclusion of 5 wt% MWCNTs were characterised by using tensile tests up to 40% strain, which were conducted parallel to the loading and transverse directions with the simultaneous measurements of sensor resistance. It was found that the resistivity isotropy was maintained during the deformation [5]. Xiang et al. [6] fabricated PU/MWCNT/sliver nanoparticle (AgNP) composites used as elastic sensors with the aid of FDM as well. Apparently, the incorporation of nanofillers improved material printability and sensitivity.

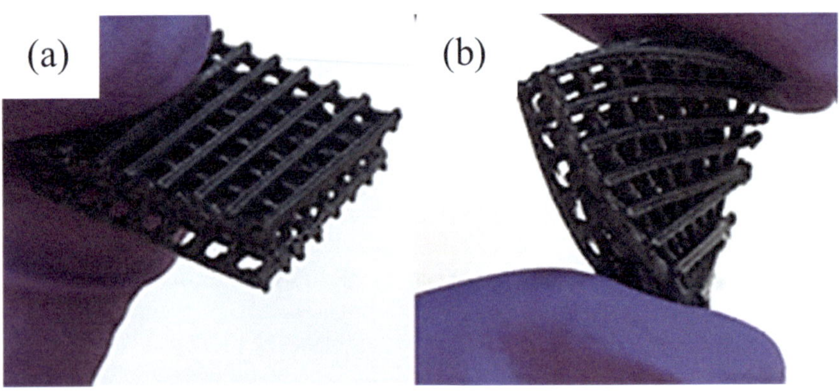

Fig. 11.1 3D printed PU/PLA/GO nanocomposite scaffold **a** without bending and **b** under bending [3]

Additionally, Fig. 11.2 displayed electromechanical reaction of such nanocomposite sensors, which resulted in rapid reaction for only approximately 57 ms, excellent sensitivity with a gauge factor (GF) of 43,260 at 250% strain, high repeatability up to 1000 cycles, as well as good linearity ($R^2 = 0.97$ within 50% strain) due to synergistic effect [6]. As such, there is a huge potential for these sensors used in wearable devices and prosthetics with typically customised requirements.

The combination of PU and PLA leads to desired properties including biocompatibility, fixity and flexibility, as per requirement by wearable devices used in biomedical engineering such as skeletal muscle actuators and finger orthosis [7, 8]. Tao et al. [8] fabricated PU/PLA finger orthosis using FDM. It was revealed that the specimens with 25 wt% PU and 75 wt% PLA possessed comparable tensile strength, but much higher elongation at break when compared with neat PLA, which was deemed the material requirement of finger orthosis. A further increase in PU content up to 50 wt% caused a drastic decrease with respect to tensile strength of such material specimens [8].

A skeletal muscle actuator is very useful to bolster motion and force to the joints within prosthetic devices. Upper limb prosthetic comprises the rotation at the elbow, digit/wrist manipulation and grip force. Hence, an actuator is required to provide a translation of rotational energy or linear output to a linear force/orientation. Nowadays, a body-powered system or an electric system has been employed in prosthetics

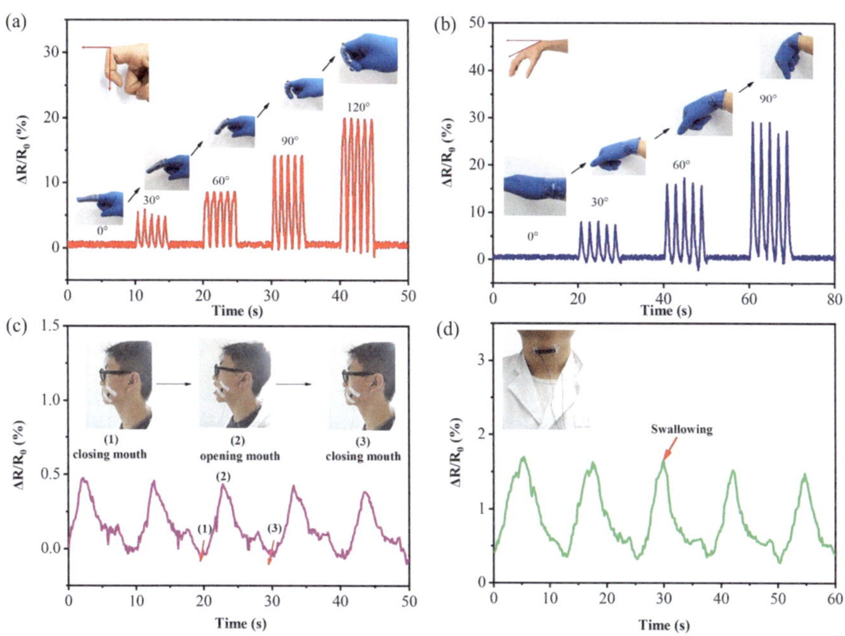

Fig. 11.2 Electromechanical sensors of PU/MWCNT/AgNP composites for repeated bending and unbending according to **a** index finger, **b** wrist, **c** mouth opening and **d** swallowing [6]

for upper limb amputation. A body-powered system is lightweight and inexpensive in lack of complexity despite no high force output or a feedback system. On the contrary, an electric system provides a feedback system, enabling to collect the inputs from muscular monitoring electrodes. Its drawback lies in high cost and heavy weight with noise pollution [7, 9, 10].

To tackle above-mentioned problems, 3D printed PU/PLA composite actuators were manufactured with the support of FDM. Since PU and PLA have a shape memory behaviour, their actuators can memorise initial form and regain it when an external stimulus such as heat is applied. 3D printed composite actuators possess a nonlinear behaviour where peak force and strain are comparable to those of mammalian skeletal muscle. Such composite actuators have been reported to gain a low mass of 74 mg, a quick response time of 0.77 s, as well as a low operating temperature of 70°C, along with a cheap price of only US$0.0098 per actuator [7]. However, loading protection within a prosthetic and durability is a major concern for these actuators. Table 11.1 lists a summary of uses and challenges for 3D printed PU composites.

11.2 Electronic Applications

3D printing technologies are popular in the field of electronic devices when processing PU composites to manufacture different components ranging from sensors, flexible circuits, thermoelectric generators, soft robots and wearable devices. Tzounis et al. [19] created flexible and stretchable organic thermoelectric devices via FDM using PU/multi-walled carbon nanotube (MWCNT) composites, consisting of NC-7000 MWCNTs (i.e. NC-MWCNT) and long MWCNTs (i.e. L-MWCNT). These devices induced thermoelectric voltage when exposed to heat and yielded the same Seebeck coefficient, but possessed anisotropic electrical conductivity in cross-layer and through-layer printing directions [19]. Such results offer a great opportunity for 3D printed thermoelectric PU/MWCNT nanocomposites to be used as stretchable thermal energy harvesters and wearable devices with customised requirements and complex geometry.

PU/MWCNT composite strain sensors to measure both uniaxial and biaxial strains were manufactured using FDM. After the investigation of their piezoresistive reaction under a series of cyclic strain loads, it was reported that such composite sensors had good reactions with cyclic repeatability in both axial and transverse directions, as well as in reaction to strains [21].

It is also worth noting in this previous study that the successful fabrication of multidirectional and embedded flexible strain sensors via 3D printing has potential applications in flexible electronics, soft robotics and health monitoring. Li et al. [20] prepared PU/CNT composite flexible strain sensors with auxetic characteristics according to Fig. 11.3. A combination of FDM and ultrasonic cavitation enabled treatment was employed to manufacture auxetic sensors with the incorporation of CNTs into their surface layers. It was concluded that PU/CNT composite sensors

Table 11.1 Applications and challenges of 3D printed PU composites in biomedical engineering [11]

Method	Materials	Application	Challenges	Ref
FDM	• PU/hexagonal boron nitride (hBN)	• Sensors	• 3D printed samples had anisotropic properties from hBN fillers	[12]
FDM	• PU/MWCNT	• Flexible sensors	• 3D printed samples inherited piezoresistivity anisotropy	[5]
FDM	• PU/MWCNT/AgNP	• Flexible sensors	• Without the presence of AgNP, CNT agglomeration in PU matrices occurred to reduce the sensitivity and linearity of 3D printed composites	[6]
FDM	• PU/PLA/GO	• Scaffolds for tissue engineering	• Printing orientation influenced different mechanical responses because of weak adhesion strength between layers during 3D printing process	[3]
FDM	• PU/PLA	• Finger orthosis	• Single-screw extruder required three times as long as a melt mixing process leading to time-consuming manufacturing	[8]
FDM	• PU/tungsten/NaCl	• Interventional radiology	• It is formidable for 3D printing PU with a low glass transition temperature T_g of 35 °C due to the loss of stiffness at room temperature	[13]
FDM	• PU waterborne/carbomethyl cellulose (CMC) • PU waterborne/silicon dioxide (SiO_2)	• Smart structures for tissue engineering	• The addition of CMC and SiO_2 decreased fracture stains of 3D printed samples • 3D printed objects still had a rough characteristic with poor accuracy • 3D printing process was time-consuming since it required 1–2 min delay between each layer	[14]
FDM	• PU/PLA	• Biomimicking skeletal muscle actuators	• 3D printed actuators had low durability • Response time was still too long	[7]
DIW	• PU/hyaluronan (HA)/transforming growth factor beta 3 (TGFβ3)	• Scaffolds	• The bulk structure of PU/HA/TGFβ3 scaffolds disintegrated after one month due to a low solid content	[15]
SLS	• PU/CNT	• Wearable devices	• Increasing CNT content decreased the tensile strength of composite samples due to incomplete coalescence of polymeric particles	[16]

(continued)

Table 11.1 (continued)

Method	Materials	Application	Challenges	Ref
DLP	• PU/polyaniline (PANI) • PU/graphene sheet (GS)	• Biomedical devices	• The effect of the addition of PANI on tensile strength and Young's modulus of composite samples was insignificant	[17]
SLA	• PU/PLA/ graphene/ triethylene glycol dimethacrylate (TEGDMA)	• Scaffolds	• Stereolithography (SLA) technique generated anisotropic behaviour on 3D printed composite structures	[18]

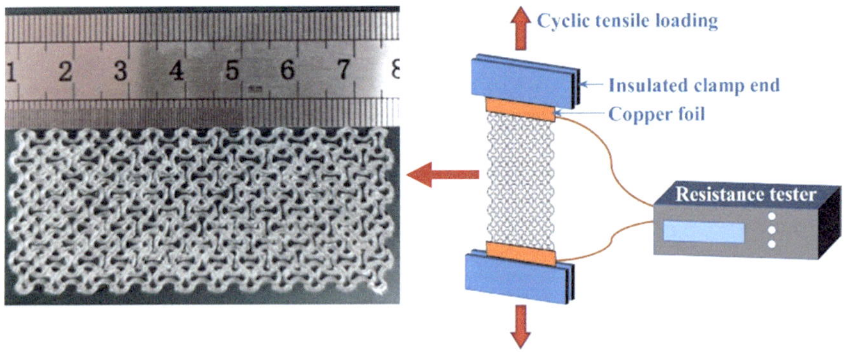

Fig. 11.3 Schematic diagram of PU/CNT composite flexible strain sensor [20]

yielded usable sensing performance with Poisson's ratio of 0.8 and the strain up to 300%. Additionally, the sensitivity within initial strain range became 2.661.

PU/carbon black (CB) composites were employed to manufacture usable sensors with remarkable electro-responsive and elastic behaviour via FDM [22]. Primary issue in sensor manufacturing was associated with enhanced brittleness of composite materials with CB inclusion. Then it sparked a premature break under tension. As such, sample temperature should be elevated to overcome such a drawback in order to pass a controlled current through the sample to adjust strength characteristics. Measurement data showed that PU/CB nanocomposite samples responded to an electric current stimulus by increasing their toughness up to 400%, which was much greater than that without the current imposed in tensile tests. However, CB inclusion in PU matrices declined the elongation at break by 50% in nanocomposite materials when compared with that of neat PU [22]. These results could be related to CB dispersion as rigid nanofillers to increase the brittleness within polymer matrices.

Biomass materials like wood flour (WF) can be considered as the reinforcements for PU matrices in PU/WF composites since they are abundant in nature with low density, low cost and environmentally friendliness [23]. The incorporation of WF as additives derived from natural resources promoted sustainable PU composites for environmental sustainability. The only drawback was chemical modifiers such as ethylene propylenediene monomer grafted with maleic anhydride (EPDM-g-MAH), which was required to be used in order to compensate for toughness loss and improve interfacial adhesion of PU/WF composites [23]. Resulting composite materials possessed remarkable tensile strength, elongation at break, as well as shape memory behaviour [23]. It was proven that WF particles could be well dispersed in PU/WF composites with the addition of 4 wt% EPDM-g-MAH. Such material composition resulted in strong matrix-filler interfacial bonding and better recovery ratio when compared with unmodified PU/WF composites [23]. Hence, PU/WF composites have a promising potential to be used in widespread applications ranging from sensors, soft robots to flexible electronics. Table 11.2 summarised 3D print applications of PU composites in electronics and their primary challenges encountered.

FDM is a more popular 3D printing technique to frequently process PU and PU composites due to its diverse material handling, low-cost manufacturing, as well as user-friendliness [25]. Typical challenges and drawbacks with respect to FDM were also listed in Table 11.1. For instance, anisotropy often emerges in a FDM process [5, 12, 19], which could be dealt with via a post-heating process. Other demerits of using FDM are not limited to filament buckling [27], material clogging and agglomeration [25]. In particular, filament buckling can be overcome by increasing filler content to improve the stiffness of filaments [24] though it may yield material jam in the nozzle for much stiffer filaments. Agglomeration issues, on the other hand, may be overcome by using an appropriate mixing method for raw materials so that used fillers enable to be uniformly dispersed in polymer matrices in a composite material system. Moreover, EPDM-g-MAH and methylene diphenyl diisocyanate (MDI) are effective modifiers to improve filler-matrix interfacial bonding [1, 3, 14, 28, 29] as a major contributor to the improvement of mechanical strength of 3D printed parts and better particle bonding in composite laminates.

The use of reinforcements may not always significantly influence material properties of unfilled polymers, as exemplified by the incorporation of PANI within PU matrices in composite materials [17]. Moreover, there is generally a threshold level of filler content, above which tensile strength may decline owing to filler aggregation and incomplete coalescence [8, 16, 26]. Hence, optimum material composition is critical to manufacture 3D printed parts with desired material properties.

11.3 Antistatic Materials

In general, antistatic materials refer to materials that can inhibit triboelectric charging. This charging means the generation of an electric charge through contact or scrubbing with other materials. When it takes place to insulating materials like plastics, electric charge usually stays in localised contact regions between the materials. The electrostatic voltage can be discharged via a spark or arc when the material contacts a body with adequately different electrical potential such as a microcircuit or a person. Once a human body encounters an electrostatic discharge, it may undergo a mild or

Table 11.2 Applications and challenges of 3D printed PU composites in electronics field [11]

Method	Material	Application	Challenge	Ref
FDM	• PU/MWCNT	• Stretchable organic thermoelectric generators	• 3D printed samples possessed anisotropic electrical conductivity	[19]
FDM	• PU/MWCNT	• Strain sensors	• The response of transverse sensors was much lower than that of axial counterparts	[21]
FDM	• PU/MWCNT	• Sensors	• A stress–strain relationship may vary significantly when repeatedly loaded up to large strain	[24]
FDM	• PU/MWCNT	• Flexible sensors	• 3D printed samples possessed piezoresistivity anisotropy	[5]
FDM	• PU/MWCNT/AgNP	• Flexible sensors	• Without the presence of AgNP, CNT agglomeration in PU matrices occurred to reduce the sensitivity and linearity of 3D printed composites	[6]
FDM	• PU/hBN	• Sensors	• 3D printed samples had anisotropic properties with the addition of hBN fillers	[12]
FDM	• PU/MWCNT	• Sensors	• Technical issues such as material clogging frequently took place so that it consumed much more time and labour	[25]
FDM	• PU/CB	• Sensors	• The incorporation of CB decreased elongation at break for composite materials up to 50%	[22]
FDM	• PU/WF/EPDM-g-MAH	• Soft robots, flexible electronics, sensors	• Increasing EPDM-g-MAH content over 4 wt% deteriorated both tensile strength and elongation at break for 3D printed samples	[23]
SLS	• PU/CNT	• Flexible circuits, wearable devices	• Increasing CNT content decreased the tensile strength of composite samples due to incomplete coalescence of polymeric particles	[16]
SLS	• PU/single-walled CNT (SWCNT)	• Flexible piezoresistive sensors	• Increasing SWCNT content undermined the tensile strength of composite samples	[26]
DLP	• PU/polyaniline (PANI) • PU/graphene sheet (GS)	• Antistatic materials, sensors	• The effect of addition of PANI on tensile strength and Young's modulus of composite samples was trivial	[17]

painful shock and even death. An electrical spark is harmful to the environment with flammable materials in solids, liquids or gases (e.g. hospital operating room). The most prevalent victims of electrostatic discharge are sensitive electronic components that can break due to this discharge. Accordingly, antistatic materials are in great demand to hinder or reduce electrostatic discharge.

Joo et al. [17] prepared PU/PANI/graphene sheet (GS) composite structures using digital light processing (DLP). It was revealed that 3D printed PU composites reinforced with 6 wt% PANI yielded the sheet resistance of 1.19×10^6 Ω/sq and the electrical conductivity of 9.28×10^{-7} S/cm. Meanwhile, the sheet resistance and electrical conductivity of PU composites reinforced with 2 wt% GS became 8.05×10^5 Ω/sq and 1.37×10^{-6} S/cm respectively [17]. On the other hand, the minimum necessity of antistatic agents and conducting pastes is based on the criterion to possess sheet resistance magnitudes of 10^{11} and 10^7 Ω/sq. It implied that PU/PANI composites and PU/GS composites were adequate to be antistatic and conducting materials with high material performance.

11.4 Thermal Insulation

Heat transfer may take place when two or more objects have distinct temperatures. Thermal insulation is required to minimise heat transfer between objects in thermal contact or radiative impact perimeter. This insulation generates thermal hindrance or break in a particular region where thermal conduction becomes minimal. The insulating ability of a material can be influenced by the inverse of its thermal conductivity, which means that low thermal conductivity has high insulating ability called a resistance value.

Gama et al. [30] built 3D printed PU/cork foams via FDM. The presence of cork and voids lowered thermal conductivity and density of 3D printed foams despite no effect on thermal stability of the structures to meet thermal insulation requirements in applications.

11.5 Sports Equipment

Sports equipment such as golf putters, rackets, bicycles, sports glasses and helmets typically requires used materials with good characteristics including high ductility, light weight, high strength, great toughness and easy customisation. These multifunctionalities can be achieved with the use of 3D printed composite materials. PU and paraffin can also be used jointly as FDM feedstock materials to fabricate composite structures with the capability of thermal energy storage (TES) [31], which refers to a special technology based on thermal energy storage in excess and release when necessary. TES can balance peak energy requirements, improve the efficiency of thermal energy management and reduce energy consumption and pollution [32]. TES systems typically consist of three main mechanisms: (i) latent heat storage which depends on the capability of storing and releasing thermal energy by gathering and releasing latent heat during a phase change; (ii) sensible heat storage which relies on cooling and heating a storage medium in liquid or solid form; (iii) thermochemical heat storage, which focuses on the storage and release of thermal energy through chemical reactions [33]. 3D printed PU/paraffin composite samples have an effective energy storage/release ability that can be used for winter sports equipment.

11.6 Garments

PU with a particular behaviour between rubbers and plastics is used as leather fabrics in the presence of moisture permeability, bionic and leather texture, as well as water and abrasive resistance. In addition, PU films are widely implemented in clothing industries for women underwear, raincoats, waterproof, wetsuit and moistureproof clothing [34].

Kabir et al. [28] used PU and nylon to produce sinusoidal sample/nylon composites via FDM. Tensile properties and shape memory behaviour of those samples were analysed with various cycles ranging from 1, 10, 20, 30, 40 to 50 and the thicknesses of 0.2, 0.4, 0.6, 0.8 and 1.0 mm. 3D printed composite samples had a shape recovery ratio of 100% for different thicknesses up to 50 cycles. The greater sample thickness induced the shape recovery rate to decline from 3 to 0.7°/s. Tensile properties, comprising tensile modulus and strength, were improved by increasing the cycle number. It has been shown that PU/nylon composites reveal great potential for smart garments, smart textiles and protective clothing.

11.7 Summary

A variety of potential applications for 3D printed PU composite structures were discussed in this chapter including biomedical engineering, electronic and antistatic devices, thermal insulation, sports equipment and garment. In biomedical fields, scaffolds, flexible sensors, finger orthosis, smart structures for tissue engineering, biomimicking skeletal muscle actuators, as well as wearable and biomedical devices, have been covered to overcome some critical issues related to human health. In electronic fields, PU composites can be used to fabricate thermoelectric generators, strain and flexible sensors, soft robots and flexible circuits. Other useful structures, such as antistatic materials, thermal insulation, sports equipment and even garments, are created based on 3D printed PU composites. It is anticipated that 3D printed PU composites have great potential to effectively produce multifunctional parts for widespread applications in contemporary advanced manufacturing.

References

1. Guerado, E., Caso, E.: Challenges of bone tissue engineering in orthopaedic patients. World J. Orthopedics **8**, 87–98 (2017)
2. Gupta, A., Maharjan, A., Kim, B.: Shape memory polyurethane and its composites for various applications. Appl. Sci. **9**, 1–29 (2019)
3. Chen, Q., Mangadlao, J., Wallat, J., Leon, A., Pokorski, J., Advincula, R.: 3D printing biocompatible polyurethane/poly(lactic acid)/graphene oxide nanocomposites: anisotropic properties. ACS Appl. Mater. Interf. **9**, 4015–4023 (2017)
4. Zhang, Y., Hu, J., Zhao, X., Xie, R., Qin, T., Ji, F.: Mechanically robust shape memory polyurethane nanocomposites for minimally invasive bone repair. ACS Appl. Bio Mater. **2**, 1056–1065 (2019)
5. Kim, M., Jung, J., Jung, S., Moon, Y., Kim, D., Kim, J.: Piezoresistive behaviour of additively manufactured multi-walled carbon nanotube/thermoplastic polyurethane Nanocomposites. Materials **12**, 1–24 (2019)
6. Xiang, D., Zhang, X., Jones, E., Zhu, W., Zhou, Z., Shen, Y., Li, Y., Zhao, C., Wang, P.: Synergistic effects of hybrid conductive nanofillers on the performance of 3D printed highly elastic strain sensors. Compos. Part A: Appl. Sci. Manufact. **129**, 1–38 (2020)
7. Lathers, S., Belle, J.: Additive manufactured biomimicking actuator with shape memory polymer composite for prosthetic actuators. In: 3D Printing and Additive Manufacturing, vol. 4, pp. 201–213. (2017)
8. Tao, Y., Shao, J., Li, P., Shi, S.: Application of a thermoplastic polyurethane/polylactic acid composite filament for 3D-printed personalized orthosis. Mater. Tehnol. **53**, 71–76 (2019)
9. Chandler, R., Clauser, C., Mcmconville, J., Reynolds, H., Young, J.: Investigation of Inertial Properties of the Human Hand. U.S. Department of Transportation, Washington, DC (1975)
10. Tozeren, A.: Human Body Dynamics: Classical Mechanic and Human Movement. Springer, New York (2000)
11. Nugroho, W., Dong, Y., Pramanik, A., Leng, J., Ramakrishna, S.: Smart polyurethane composites for 3D or 4D printing: general-purpose use, sustainability and shape memory effect. Compos. Part B Eng. **223**, 109104 (2021)
12. Liu, J., Li, W., Guo, Y., Zhang, H., Zhang, Z.: Improved thermal conductivity of thermoplastic polyurethane via aligned boron nitride platelets assisted by 3D printing. Composite Part A **120**, 140–146 (2019)
13. Kashyap, D., Kumar, P., Kanagaraj, S.: 4D printed porous radiopaque shape memory polyurethane for endovascular embolization. Additive Manufact. **24**, 687–695 (2018)
14. Su, W., Gao, W., Trinh, K., Kenderes, S., Pulatsu, E., Zhang, C., Whittington, A., Lin, M., Lin, J.: 4D printing of polyurethane paint-based composites. Int. J. Smart Nano Mater. **10**, 237–248 (2019)
15. Hung, C., Tseng, C., Dai, L., Hsu, S.: Water-based polyurethane 3D printed scaffolds with controlled release function for customized cartilage tissue engineering. Biomaterials **83**, 156–168 (2016)
16. Li, Z., Wang, Z., Gan, X., Fu, D., Fei, G., Xia, H.: Selective laser sintering 3D printing: a way to construct 3D electrically conductive segregated network in polymer matrix. Macromolecular Mater. Eng. **1700211**, 1–10 (2017)
17. Joo, H., Cho, S.: Comparative studies on polyurethane composites filled with polyaniline and graphene for DLP-type 3D printing. Polymers **12**, 67 (2020)
18. Feng, Z., Li, Y., Hao, L., Yang, Y., Tang, T., Tang, D., Xiong, W.: Graphene-reinforced biodegradable resin composites for stereolithographic 3D printing of bone structure scaffolds. Hindawi J. Nanomater. **2019**, 1–14 (2019)
19. Tzounis, L., Petousis, M., Grammatikos, S., Vidakis, N.: 3D printed thermoelectric polyurethane/multiwalled carbon nanotube nanocomposites: a novel approach towards the fabrication of flexible and stretchable organic thermoelectrics. Materials **13**, 1–16 (2020)

20. Li, B., Liang, W., Zhang, L., Ren, F., Xuan, F.: TPU/CNTs flexible strain sensor with auxetic structure via a novel hybrid manufacturing process of fused deposition modeling 3D printing and ultrasonic cavitation-enabled treatment. Sens. Actuators A: Phys. **340**, 113526 (2022)
21. Christ, J., Hohimer, C., Aliheidari, N., Ameli, A., Mo, C., Potschke, P. 3D printing of highly elastic strain sensors using polyurethane/multiwall carbon nonotube composites. In: Paper presented at Proceeding SPIE 10168, Sensors and Smart Structures Technologies for Civil, Mechanical, and Aerospace Systems, vol. 101680E, pp. 1–7 (2017)
22. Rosales, C., Duarte, M., Kim, H., Chavez, L., Hodges, D., Mandal, P., Lin, Y., Tseng, T.: 3D printing of shape memory polymer (SMP)/carbon black (CB) nanocomposites with electro-responsive toughness enhancement. Mater. Res. Expr. **5**, 1–14 (2018)
23. Bi, H., Ren, Z., Guo, R., Xu, M., Song, Y.: Fabrication of flexible wood flour/thermoplastic polyurethane elastomer composites using fused deposition molding. Indus. Crops Prod. **122**, 76–84 (2018)
24. Xiang, D., Zhang, X., Li, Y., Jones, E., Zheng, Y., Wang, L., Zhao, C., Wang, P.: Enhanced performance of 3D printed highly elastic strain sensors of carbon nanotube/thermoplastic polyurethane nanocomposites via non-covalent interactions. Compos. Part B Eng. **176**, 107250 (2019)
25. Ly, S., Kim, J.: 4D Printing—Fused deposition modeling printing with thermal-responsive shape memory polymers. Int. J. Precis. Eng. Manufact. Green Technol. **4**, 267–272 (2017)
26. Gan, X., Wang, J., Wang, Z., Zheng, Z., Lavorgna, M., Ronca, A., Fei, G., Xia, H.: Simultaneous realization of conductive segregation network microstructure and minimal surface porous macrostructure by SLS 3D printing. Mater. Design **178**, 1–10 (2019)
27. Yao, X., Qi, X., He, Y., Tan, D., Chen, F., Fu, Q.: Simultaneous reinforcing and toughening of polyurethane via grafting on the surface of microfibrillated cellulose. J. ACS Appl. Mater. Interf. **6**, 2497–2507 (2014)
28. Kabir, S., Lee, S.: Study of shape memory and tensile property of 3D printed sinusoidal sample/nylon composite focused on various thicknesses and shape memory cycles. Polymers **12**, 1–17 (2020)
29. Jing, X., Mi, H., Peng, X., Turng, L.: The morphology, properties, and shape memory behavior of polylactic acid/thermoplastic polyurethane blends. Polym. Eng. Sci. **55**, 70–80 (2014)
30. Gama, N., Ferreira, A., Timmons, A.: 3D printed cork/polyurethane composite foams. Mater. Design **179**, 1–9 (2019)
31. Rigotti, D., Dorigato, A., Pegoretti, A.: 3D printable thermoplastic polyurethane blends with thermal energy storage/release capabilities. Mater. Today Commun. **15**, 228–235 (2018)
32. International Energy Agency-The Energy Technology System Analysis Programme (IEA-ETSAP).: Energy supply technologies data (2013). https://iea-etsap.org/index.php/energy-technology-data/energy-supply-technologies-data
33. Pielichowska, K., Pielichowski, K.: Phase change materials for thermal energy storage. Progress Mater. Sci. **65**, 67–123 (2014)
34. Wang, Y., Li, X., Chen, Y.: Research on innovation and application of thermoplastic polyurethane elastomer bionic leather fabric from the perspective of sustainable design. In: Paper presented at IOP Conf. Series: Earth and Environmental Science, vol. 706, pp. 012032. (2021)

Correction to: Influence of HNT Content and 3D Printing Parameters on Shape Memory Behaviour and Compressive Properties of PU/HNT Nanocomposites

Correction to:
Chapter 10 in: W. T. Nugroho et al., *Thermoplastic Polyurethane (TPU)/Halloysite Nanotube (HNT) Nanocomposites*, Engineering Materials, https://doi.org/10.1007/978-981-95-0541-8_10

In the original version of the book, the following belated correction has been incorporated: In Chapter 10, incorrect figure (figure 10.4) was updated which has now been replaced with a similar image. The book has been updated with the changes.

The updated version of this chapter can be found at
https://doi.org/10.1007/978-981-95-0541-8_10

W. T. Nugroho et al., *Thermoplastic Polyurethane (TPU)/Halloysite Nanotube (HNT) Nanocomposites*, Engineering Materials, https://doi.org/10.1007/978-981-95-0541-8_12

MIX
Papier aus verantwortungsvollen Quellen
Paper from responsible sources
FSC® C105338

If you have any concerns about our products,
you can contact us on
ProductSafety@springernature.com

In case Publisher is established outside the EU,
the EU authorized representative is:
Springer Nature Customer Service Center GmbH
Europaplatz 3, 69115 Heidelberg, Germany

Printed by Libri Plureos GmbH
in Hamburg, Germany